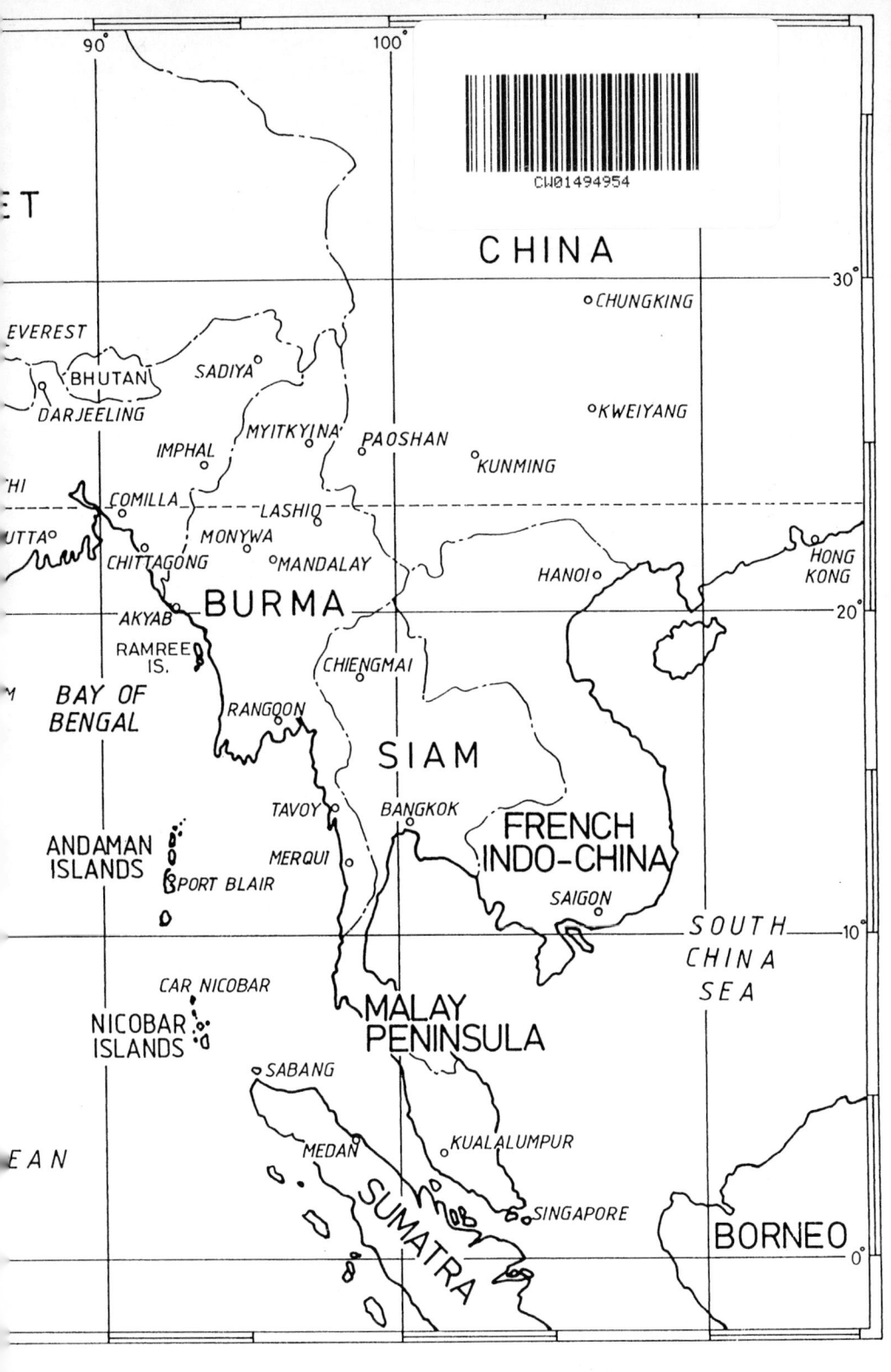

To Burma Skies and Beyond
An Airman's Story

To Burma Skies and Beyond
An Airman's Story

Group Captain Reg (Lucky) Jordan DFC AFC

JANUS PUBLISHING COMPANY
London, England

First published in Great Britain 1995
by Janus Publishing Company
Edinburgh House, 19 Nassau Street
London W1N 7RE

Copyright © Group Captain Reg Jordan

British Library Cataloguing-in-Publication Data.
A catalogue record for this book is available
from the British Library.

ISBN 1 85756 138 4

Printed & bound in England by
Antony Rowe Ltd,
Chippenham, Wiltshire

Contents

Part Three – Tailpiece

FOREWORD
by
Air Chief Marshal Sir John Aiken KCB
Former Air Member for Personnel

I have known Reginald Jordan for some 35 years. He has written this personal story of his life in aviation, the greater part of which covers his service in the Royal Air Force.

His account of his entry into the R.A.F. as aircrew, having been an enthusiastic aviation teenager in the 1930s, describes well the aspirations of many of his generation. It is evocative of a period which saw a huge surge in the build-up of this service to meet the demands of war.

At a time when, fifty years after the end of World War II, there is a focus of attention upon the operations which took place in the various theatres, his description of his tour as a Liberator captain in support of the Burma Campaign makes interesting reading. How he came to be there, the length of some of the sorties and the constraints upon operational effectiveness should be of value to historians of the use of air power in that region.

Lastly, when he returned to the R.A.F. in 1949, becoming a qualified flying instructor, he joined a band of former operational and experienced pilots who made a major contribution to flying training in the Royal Air Force. It also extended to other countries whose air forces used our facilities. They built upon the achievements of the Central Flying School, the oldest unit in the Royal Air Force, in the 1920s and 1930s and laid a new foundation which has lasted until the present time.

ACKNOWLEDGEMENTS

I wish to acknowledge the help of the Public Record Office at Kew in giving me access to the RAF 231 Group Headquarters and 356 Squadron diaries covering the Burma Campaign, and of those friends who kindly read the manuscript during its compilation and encouraged its publication. In particular, I wish to thank Harry Holmes, the Aviation and Naval Historian, for his ready and helpful advice in guiding me through the pre-publication stage. My friends Group Captain Charles ('Ace') Newman and Alan Troughton also deserve my thanks for their assistance in identifying the market for my book. I must express my gratitude too to David Fildes, who produced the painting on the cover, for bearing with me in my pursuit of accuracy. Finally, I am indebted to Air Chief Marshal Sir John Aiken and to Frank Gillard who gave so generously of their time in wading through the pages at the proof stage.

Part One
To Burma Skies

1   *Enthusiasm and Luck*

Hooked

There was a general gasp as it came into view its silver shape glistening enticingly in the afternoon sunshine, like some huge dragonfly which had settled to dry its wings. Our quarry in sight, we went at a jog the rest of the way; arriving panting and with the stitch.

The year was 1932. I was rising nine years old and, whilst in most respects a pretty ordinary boy, I was mad about aircraft. In our rural surroundings the trouble lay in getting anywhere near to one. So when news reached our school that an aircraft had been forced to land a mere mile away, I could barely wait for lessons to end and to be in search of it. I was not alone.

Our hearts still thumping, we padded round and round that beautiful creature: touching it now and then; noting that with its canvas skin and profusion of bracing wires it looked rather frail as well as being remarkably elegant; admiring the craftsmanship which had so lovingly shaped its wooden propeller – a harmony of polished laminations and graceful curves; gazing inquisitively at the lifeless engine; questioning the men with it; and in between, just standing enthralled. Had it been a machine from outer space, it could scarcely have commanded keener attention from our eager band.

Shortly afterwards, aviation made a more deliberate visit to our town (Wellington in Somerset) this time in the shape of Mr Cathcart Jones's flying circus. Not such a household name as that of Sir Alan Cobham perhaps, but we flocked to the field at Chelston none the less. My friend Melvin brought his mother along, and she was sufficiently carried away to offer him the price of a joy ride. But Melvin was happier with one foot on the ground. And although, for the price of a ticket, I would gladly have kept him company, his mother chose to ignore my enquiring glances.

The joy-ride market exhausted, Mr Jones got down to the real business of the day. Suddenly the leisurely, confidence-building pace was replaced by a pulsating whizz-bang tempo, as one noisy, colourful spectacle gave way to another. For the next forty minutes it was manna: formation flying; aerobatics; a series of attacks on a mock fort, which obligingly emitted explosions and belched flame and smoke – not always on cue; an old banger being dive-bombed with bags of flour as it zig-zagged frantically only yards from where we stood; and huge bunches of toy balloons being torn to shreds by the propellers of marauding aircraft. Only one thing detracted from a feeling of utter contentment as I strolled home that evening – my souvenir programme clutched lovingly in my hand – that missed opportunity of Melvin's.

As I entered my teens, aviation was bursting out like buds in spring: a fact not lost on the publishing world, nor, for that matter, our local branch of W H Smith. We up-and-coming aviators deserved and got a publication to ourselves. For a modest threepence, the *Flying Weekly* told it all: whether it was the first flight of an RAF prototype, the latest activities of the newly formed Civil Air Guard or providing its readership with drawings from which they could construct an aircraft of their very own (the ill-fated Flying Flea). One week I went completely overboard and spent three shillings and sixpence on a special edition of the *Illustrated London News* devoted entirely to the expansion of the RAF: a small fortune, but somehow I just had to have that edition; its handsome blue and gold covers in no way over-stating the contents. On the contrary, it was brimming with photographs, sketches and exploded diagrams of the latest aircraft: Battles, Blenheims, Hampdens, Hurricanes – you name them, they were there. By the time I approached my fifteenth birthday, I was hooked.

My friends and I soaked up aircraft particulars with the facility of a dry sponge; civil and military, RAF and foreign. If the average adult wanted to know about aircraft he consulted one of us, not another adult: and the same went for aircraft recognition. It was practically a point of honour that no aircraft should pass overhead without being identified, and the slightest drone from the sky had me dashing from the house. Too often those excursions yielded nothing more exciting than the benign Avro Anson (my home town being about as far removed from the RAF's operational bases as was possible). But at the time of the Munich crisis we enjoyed a purple patch when the RAF showed the flag with substantial formations of front-line aircraft, even over dozy Somerset: some of them sadly antiquated, such was our lack of preparedness.

And although those flag-flying formations were one-off affairs, from Munich onwards – in terms of individual aircraft at least – the RAF began to show it face more purposefully. Low flying in the vicinity of Fox Brothers' Tonedale woollen factory, where I worked as the lodge keeper's assistant, became a daily occurrence. Fairey Battles and Hawker Hurricanes almost clipping the factory rooftops as they followed the railway towards Exeter. Old Hugh Fox, the company chairman, found this behaviour most disturbing, and I was instructed to take down the registration details so that the pilots could be reported. But with more empathy for the pilots than with Mr Hugh's concern for our safety, I'm afraid I made a poor witness for the prosecution.

In the summer of 1939, along with a fellow enthusiast, I joined the Taunton Wing of the Air Defence Cadet Corps (ADCC). Attending the weekly parade meant cycling the eight miles out and back at the end of the working day. But with the promise that the average cadet could expect to make sergeant pilot before he was twenty, the last thing we lacked was motivation.

Regrettably, at that time the ADCC training syllabus had precious little to do with flying. Most of our time being spent on rifle drill; punctuated now and then by a lecture on other small arms, like the Thompson sub-machine-gun, the latter a shade more interesting than the Lee Enfield rifle perhaps, but still dull stuff if you're an aspiring sergeant pilot. If it had not been for an evening's practical instruction in motorcycling – which really had my adrenaline pump-

ing – I'm afraid I would have to write off my membership of the
ADCC as a non-event. Better was to come from the Air Training
Corps (ATC), the successor organisation.

It's an Ill Wind . . .

Promptly on my seventeenth birthday I volunteered for training
as an RAF Wireless Operator/Air Gunner, attending the Exeter
Recruiting Office for a preliminary assessment. I came away with
the feeling of having lost sixpence and found half a crown – meanwhile having been encouraged to raise my sights to the pilot/navi-
gator category. The rider that I would need to work on my maths
seemed a small price to pay.

It was September 1940; the phoney war was behind us; to gain
the initiative the RAF needed to expand and to do so rapidly. Finding the aircrew numbers required meant casting the net wider,
drawing on people like myself and raising our educational standards
by pre-entry training. And, although I was unaware of it at that
time, the situation also called for some lowering of medical standards.
'It's an ill wind . . .'

My formal aircrew selection interview took place in January 1941
at the Clarendon Laboratories in Oxford. The board having decided
I was suitable for pilot/navigator training, the president's parting
shot was to wish me luck in defeating the medicos: I was to need it.

All went well until I was invited to look through *that* book – the
one containing page after page of coloured dots. My problem was
that a percentage of the characters which I could make out on those
devilish pages, would be invisible to someone with normal colour
vision, and vice versa. Technically speaking, I was red/green blind!
Another test followed, where I was placed in a darkened room with
a small light on the far wall. For about ten minutes the size and
colour of that light was varied as the medico gauged the real extent
of my colour blindness.

After some hesitancy, the medical officer – young, and possibly
new to the game – pronounced my colour vision to be 'defective-
safe'. He did suggest, however, that on operations it might be sensible

to have my navigator confirm the colour of any signal which was flashed to me. Without wishing to appear ungrateful, I asked, 'But what if I'm on single-engined fighters?' to which he replied, rather tartly, 'Look, I'm trying to help you. You want to be a pilot, don't you?' I did not push my luck.

The medics not so much defeated as having actively come to my aid, I was formerly attested and released on deferred service pending my eighteenth birthday. What luck indeed: the lowering of the medical standards and my being handled by that particular – flexibly-minded – medical officer.

At Oxford they also gave me a maths syllabus; a formidable document detailing the heights to be scaled before call-up. A lot of work lay ahead, and I set about canvassing assistance. Looking back, I can scarcely credit the amount of help already busy people were prepared to give me once they sensed my degree of commitment: Andrew Brunton my employer's industrial chemist, for instance, who corrected my maths when he might have been on the golf course, and who also gave me a blast if my work showed any sloppiness.

Early in 1941, Wellington got its own ATC squadron – No 357. Douglas Tancock, one of my former school teachers, was the driving force behind its formation and enlisted me as one of his first cadets. There is nothing like being in on the ground floor of a new organisation. At our second parade I was given the rank of Corporal, and further promotion followed at a dazzling pace as the rank structure of a complete squadron was fleshed out. In a matter of weeks, it seemed, I was a Flight Sergeant. Pleased as he was, my father could scarcely credit my rate of advancement.

Our ATC training put the emphasis where it belonged – on things related to flying. In our navigation classes we added rhumb lines and great circles to our vocabularies and wrestled with the triangle of velocities; we listened to the Morse code until our very heads buzzed, although we never got much beyond six words per minute; we put a sharp edge on our powers of aircraft recognition; and the drill practice helped reduce the trauma of the 'square bashing' which greeted our entry into the RAF proper. Yes, thanks to our officers

– every one of them with a full-time job to cope with – the ATC
did a lot to lick us into shape.

Not that the ATC was awash with training equipment; a few
manuals apart. But goodwill on the part of other organisations and
our own ingenuity helped fill the gaps. Wellington School had a
Morse buzzer, so we went along there for tuition in Morse: and very
effective tuition it was, delivered personally by Aubrey Price, the
headmaster (ex 1914–1918 Royal Flying Corps) with great aplomb,
and full of ingenious ways of associating sounds with letters. Very
soon 'dit D A H dit dit' (the Morse sound of the letter L) came to
be heard as 'Like E L L e is'; 'dit dit D A H dit' (F) as 'Fun ny F
E L L ow', and so on.

The local Observer Corps also came to our assistance by inviting
us to one of their evening training sessions. Harry Croft, who
conducted these sessions in the attic of his ironmonger's shop, had
made his own episcope for projecting aircraft silhouettes, and the
sight of this in operation, fired me up to produce one for the squadron. That same night Harry generously parted with the drawings,
and a fortnight later we had one too: the glass from my father's old
carbide bicycle lamp making an excellent lens, and my carpentering
uncle came up with a cabinet to house the lamp bulbs and reflectors.
Admittedly our contraption smelt a bit once it warmed up, but it
did an excellent job.

Now and then our lecture periods were given a special zest by
the attendance of a real, practicing aviator. One evening the task
falling to Howard Kallend, an aircrew cadet, whose chatty account
of life as a trainee pilot was received with something approaching
awe. Howard, whose father kept a fish shop in the town's High
Street, was so polished and debonair that I suspect most of us felt
those social attributes might be just as necessary to success as mastering the academic syllabus. If the intention had been in any way to
head off over-confidence on our part, then Howard's panache certainly took care of that.

On another occasion, the radio broadcaster Frank Gillard (also a
Wellingtonian) came along to a training session. In those days I
believe he was still a school teacher and only in an advanced courtship

with the BBC. Later, of course, he was to distinguish himself as a BBC War Correspondent, and in postwar days as a managing director in BBC radio. It was no social visit. He arrived with a van load of recording apparatus, and with the assistance of his engineers proceeded to record our activities. Broadcast shortly afterwards on the Home Service, and placing 357 Squadron squarely on the map, it was heady stuff.

2   The RAF – Early Impressions

Call-Up

By October 1941, a month had elapsed since my eighteenth birthday, and still no call-up. I wrote reminding the authorities of my attestation the previous January. Back came a terse assurance that I had not been overlooked. But it was late December before my papers finally arrived – I was to report to the Aircrew Receiving Centre (ACRC) at Lord's cricket ground on 19th of January.

With the requirement to arrive by ten in the morning, I stayed overnight in a Salvation Army hostel close to Paddington station. Sleep did not come easily. The room was packed with twin-tiered bunks, and people trickled in until the small hours. To complicate matters, the last man in left the lights on, which was how they remained for the rest of the night. In my bemused state, I half wondered if he was merely complying with hostel regulations.

The next day I was firmly caught up in the ACRC induction system along with hundreds of other aspiring pilots and navigators. We were numbered, photographed, checked for venereal disease, given a kitbag, a uniform, a button stick and – encouragingly – a set of cutlery, a mug and plate. And whilst all this was going on it gradually dawned on us that the RAF had a language of its own, at least at barrack-room level. A colourful, often mocking one: whereby our cutlery became our 'irons' and being 'put on a fizzer' meant being charged with a disciplinary offence. It was a language with a vaguely anarchic flavour; a useful safety valve for the old sweats, no doubt; and for new recruits like ourselves, something novel to indulge in.

Already, on that first day, wags emerged from among our intake to make light of having to provide details of our next of kin; being issued with *imperishable* identity discs; receiving our first inoculations; and being examined for the pox. By that evening, even the calculated abrasiveness of our corporal-in-charge was a source of amusement, as he was beautifully mimed by one of our brighter sparks. And as the days went by, this bond of shared misfortune seemed to weld us together; something which did a lot to soften the rough edges of the system. It was a feature I much enjoyed, and stronger than anything I could recall from my school-days.

Those of us who had been attested for more than six months had to undergo a complete re-run of the aircrew medical examination. Again I failed the test involving the book of coloured dots; again this was followed by the more practical test using coloured lights; and again, I got through. On this occasion I was able to observe the light as the man ahead of me was tested, and since I could also hear his responses and he got a pass score, I had a useful basis from which to work. As I finished my test, the medico exclaimed with some exasperation, 'Why on earth are they wasting our time sending people like you for testing? Your colour vision is perfectly all right!' I was glad he thought so, and I hoped that when it came to the real thing I would not make some dreadful mistake.

There were other tests too. About fifty per cent of us failed the one on maths, including me, and a day or so later we failures were installed in the Metropole hotel at Brighton receiving extra tuition.

The business of brushing up our maths was interspersed with copious amounts of drill, occasionally with gas masks fitted. So drill periods at Brighton became something of a bind, in spite of that invigorating sea air. But authority did not always rule the day. One afternoon, as our drill Corporal led us out – yet again – between the Metropole and the Forte's café directly alongside, the two rear files, to a man, wheeled smartly right and into the café. Safely ensconced, with the flight disappearing towards Hove, we indulged ourselves in toasted tea cakes and much nervous laughter before sneaking back to our rooms.

The only other way in which I cocked a snook at authority at that time was to accompany a fellow cadet on a weekend visit to his parents at Croydon when technically our course was confined to 'camp'. Both crimes remaining undetected, I'm happy to say. But dodging the service police at the mainline stations on that Croydon escapade, although providing a sense of achievement in retrospect, found us pretty tense at the time.

In spite of that bit of bolshiness involving the café, when we *were* being drilled we gave it all we had, marching at one hundred and forty paces per minutes, and swinging our arms to shoulder height. What is more, in short bursts it could be quite stimulating. Certainly, it looked smart performed by aircrew cadets, most of us in our teens and eager to impress. But, sad to say, there were some receiving drill on the Brighton sea front, whose appearance and performance was anything but inspiring. They were RAF officers, among them Wing Commanders and Squadron Leaders, and word had it that they were being disciplined for refusing to fly on operations. Some of them wore scruffy, faded uniforms and, as a squad, they looked a tired, beaten lot. Through the grapevine, we were aware that aircrew who refused to fly on operations were classified as 'Lacking in Moral Fibre' and that their service records bore the letters LMF: so that wherever they went, the odium remained with them. But, presumably, those broken-looking officers had started out like us, full of idealism and ambition. It was not a thought to dwell on.

Initial Training
Our maths up to standard, we began our pre-flying training in earnest. For this we were despatched to one of several Initial Training Wings (ITWs) dotted around the coast, in my case No 6 ITW at Aberystwyth. Whether the idea was in part to remove us from the target areas of enemy bombers, I can't say, but those seaside locations certainly provided a healthy physical environment. It was early March, and the wind whipping in off Cardigan Bay was nothing if not bracing.

As for our drill instructors, they must have felt they were in paradise, with those acres of promenade to play with. Too many acres it seemed, if we were to believe a report from one of the ITWs. According to this, the leading files of a flight getting out of earshot and wearing respirators which had misted up, marched clean over the edge of the promenade and finished up in a heap on the beach below. True or not, it made a good story. And with promenade railings removed in the drive for scrap metal, it seemed just possible.

We were joined at Aberystwyth by a small number of regular (pre-war) airmen: Corporals and Leading Aircraftmen who, like us, had volunteered for aircrew training. The majority of them had enlisted as apprentices and were qualified tradesmen, and no doubt it was something of a test of their adaptability to be pitched in alongside newly joined eighteen year olds like myself. But generally they managed it remarkably well, even if now and then they could not resist reminding us that they had seen it all before. Undoubtedly we matured more quickly for their presence as well as benefitting from their know-how.

We were accommodated in the Queen's Hotel; like the Metropole at Brighton, situated on the sea front. The nearest public house was the Boar's Head: immediately christened the Whore's Bed. Not that we frequented it: that establishment, like most of the Aberystwyth pubs, being out of bounds to the ordinary airman.

The ex-apprentices' barrack-room poetry, although larded with obscenities, had many a splendid turn of phrase, and one could half believe the rumours that it had been written by such celebrated manipulators of the language as Noël Coward. Most of these pieces, 'Eskimo Nell' for instance, had umpteen verses, and our seasoned mates had them off to a T. Among the other accomplishments which distinguished the regular airman, was his ability to gulp down huge quantities of air with which to belch titles and whole phrases, usually in mock disrespect for authority. The hallmark of an accomplished belcher being his capacity to deliver, unimpeded, 'The Archbishop of Canterbury'; with 'Works and Bricks' thrown in by way of counterpoint.

The dread of being mistaken for a new recruit was quite marked in some of the regulars, and for a few it assumed cult-like proportions. On drawing a new uniform, they would begin by wiping the floor with it in order to age it more quickly; this ageing process being helped along by shaving down the surface nap to simulate years of wear. Status symbols come in odd guises.

The presence or absence of the small insignia immediately below the eagles on our jacket sleeves, was also made much of by the cult. The complete absence of these small insignia indicated that the wearer was indeed a regular airman: the top of the pile. Next came members of the Auxiliary Air Force (also a wholly pre-war entity, but whose members served only on a part-time basis until hostilities were judged to be imminent), whose jackets bore the letter 'A'. And at the bottom of the heap came the likes of myself, members of the Volunteer Reserve, with the insignia 'VR', and who had joined after the outbreak of war and only 'for the duration of the present emergency', to quote the jargon.

These niceties of status, however, were largely forgotten as we got to know one another, and as the academic syllabus increasingly absorbed our attention. There were dry subjects, like the running of the NAAFI and the ramifications of Air Force Law; disturbing ones, like the effects of gas warfare and venereal disease; subjects demanding intense short-term concentration, like the Morse code; technical subjects like the operation of the internal combustion engine, where great stress was laid on the mixture in the cylinder *expanding* rather than *exploding* on ignition; and, more to the point, subjects directly related to the business of flying, such as navigation and meteorology which, together with an occasional aircraft recognition test, provided the yeast which sustained our interest and motivation.

Neither was our physical well-being neglected. Frequent periods of PT were punctuated by the occasional cross-country run, and our drill sergeant more than earned his keep. The result was that we kept in good physical shape, despite an intimate association with classroom chairs; a fact proved to my satisfaction during our mid-course long weekend. Arriving at Taunton railway station close on one-thirty in the morning with Frank Discombe (a fellow cadet at Aberystwyth and also a Wellingtonian) and failing to raise a taxi, Frank and I took the nine miles walk home more or less in our stride – side pack, respirator and all. It was about four o'clock when my parents responded to a handful of gravel rattling against their bedroom window, and came down to let me in.

Discipline at Aberystwyth was inculcated in two ways: directly, which embraced drill periods, saluting, and the formal way in which we addressed our seniors; and indirectly, which included the daily requirement to leave our blankets and sheets, and anything else visible within our bedspaces, in a carefully contrived regulation pattern, formal inspections demanding even greater zeal. I could cheerfully put something into the direct category, since I was showing off me personally, and the requirement was both transitory and in response to the actual presence of authority. But the indirect category, whilst I intrinsically prefer order to chaos, I suffered and got on with, but with no real enthusiasm. Not so some of my colleagues, who would go to almost any lengths to impress the inspecting officer. One of our corporals even blancoed the bars on our fire grate in an effort to gain our room extra points: my own view being that this might encourage the inspecting officer to look for even more novel expressions of subservience on the next occasion. Some things I would not be sorry to leave behind at the ITW.

After the mid-course break, the passing of the weeks seemed to accelerate, and almost suddenly the end of the course was in sight. The vast majority of us worked hard for a good result in our written examinations, and we drew on hidden reserves as we marched past the Wing Commander at our final parade; delighting our drill Sergeant and even surprising ourselves with our snap and precision.

But the moment when tensions really evaporated came on the day when we crowded round the noticeboard and found we had passed our written examinations: the day when those of us who had been Aircraftmen Second Class (AC2s) stole away to adorn our tunics with the propeller badges of Leading Aircraftmen (LAC) before the end-of-course party. Life at Aberystwyth hadn't been all that bad, we told ourselves, as the tedious bits were swept to one side in the general euphoria. We had made the grade. The ITW phase was behind us, and already we were having animated discussions about the next stage of training – pilot grading.

We were not quite done, however. The end-of-course party aside, there remained our individual interviews with the Wing Commander, a small, retiring man, whom we had hardly seen during our course, and only on formal occasions at that. Lined up outside his office, it seemed clear that the impression we made in our allotted five minutes might easily swing our end-of-course assessments either way: after all, he hardly knew us as individuals. At that point I cast a covetous eye at the forage cap of the cadet three ahead of me, which sported an aircrew flash, the envy of the entire flight. Even the old sweats, with all their know-how about 'bull', could not match its pristine whiteness and crisp-looking finish.

I marched in smartly as my name was called. And as the Wing Commander looked up, he smiled approvingly as he took in that super white flash. Whether he twigged that he'd seen it only minutes earlier, it's hard to say. But if so, he let it pass. I could feel the interview going well as my responses flowed to a series of easy questions: my answers giving rein to my enthusiasm, but consciously avoiding any appearance of challenging his control. I left feeling quite encouraged, and only later, as we exchanged accounts of our interviews, did I feel that my experience sounded a good bit like those of my chums. It seemed that the real ace that day might have been the shy little man himself.

Our end-of-course party was held in one of the more distant pubs one not reserved for officers or senior NCOs and sufficiently removed from the Queen's Hotel to allow everyone to let go a bit. It was an opportunity to thank our instructors individually for persevering with us, and to buy them a drink before they set about the next intake. There was quite a bit of letting go that night, and the veterans on our course did a rescue act on more than one of us fledgling airmen, as legs and stomachs – unaccustomed to the effects of alcohol in quantity – faltered, and in some instances gave out altogether as the evening neared its end. The next morning, many of us felt far from well, and there were some remorseful recollections as well as a few sheepish apologies.

Fortunately hangovers do not last, and by the time Frank Discombe and I pulled into Taunton station a day or so later to enjoy a spot of leave, we were again as fit as fleas and really quite pleased with life.

3 Pilot Training

First Solo

And now for the acid test. Would we be any good as pilots? From the classroom desk and the drill square of Aberystwyth we went to a pilot grading school at Sywell, a small grass airfield on the outskirts of Northampton, whose role was to assess our potential over twelve hours of dual instruction.

The aircraft were DH82a Tiger Moths – fabric-covered biplanes with two open cockpits in tandem. The air conditioning being obligatory, goggles and warm clothing were a must. To communicate with the infallible one in the front cockpit, you bellowed into a voice tube. But mostly one listened; acknowledged briefly; then tried to emulate his latest piece of wizardry.

Collecting our flying clothing on the morning of arrival (a whole kitbag of goodies including fleecy-lined boots, three pairs of gloves and an inner and outer flying suit), most of us had our first flight the same day.

Strapping myself in on that first occasion, I was struck by the aircraft's sheer simplicity. Looking up at the fuel tank in the centre section of the top mainplane, the contents gauge was nothing more than a simple float in a glass column. And glancing left at the forward interplane strut, the external airspeed indicator comprised a simple spring-loaded metal plate, free under air pressure to be forced back alongside a static quadrant calibrated in miles per hour. Taking it all in, I wondered how the designer had managed to draw the line between functional simplicity and downright flimsiness, always assuming he'd got it right.

Yes, at rest, the Tiger Moth – with its taut fabric covering, its struts and wires, and curious half-doors – seemed a fragile, almost apologetic, creation. But once the painstaking pre-start drill was out of the way, and with a final swing of the propeller the engine burst into life, the Tiger was transformed into a vibrant, living thing. Taxiing out to the take-off position – pushing the nose first one way and then the other to see ahead; the tailskid rumbling softly over the grass; giving an occasional burst of throttle to keep the aircraft moving – was to become a time of pleasurable anticipation in the days which followed.

And what of that first flight, my first time aloft in an aircraft of any kind? How was that? In the RAF language of the day – *wizard*! First the exhilarating charge across the airfield, the slipstream from the propeller pinning my goggles firmly against my face. The tail coming up so that the nose no longer blocked off the view ahead. The final bumping of the undercarriage as we lifted off. The ground slipping away beneath. The expanding view as we clawed our way upwards. The wind in one's face. The busy buzz of the engine. The cars, people, buildings – all diminishing in size and taking on the appearance of toytown as we continued in the climb. By the end of that first flight, I knew the environment and I were going to get along fine. The question was – would I show the necessary skill?

The student pilot has an emphatic advantage over the learner motorist. Both he and his instructor have their own sets of controls – interlinked one with the other – so that when the instructor demonstrates a manoeuvre he, the student, can follow through with his hands and feet, getting the feel of what is required as well as taking in the instructor's description of what he is doing.

My familiarization flight behind me, we next explored the effect of the flying controls. RAF flying instruction takes nothing for granted. With the aircraft flying straight and level, I was asked to place my right hand lightly on the control column – the 'stick' – and as my instructor eased it gently forward, he pointed out that the nose moved in sympathy. When he pulled it back, again the nose moved in the same direction as the stick; stick to the right and the right wing went down; and so on. I was then invited to place my feet on the rudder bar and to note that when rudder was applied the nose swung in the same direction.

Next, I was told to pick a point straight ahead on the horizon and, taking control of the rudder bar, to maintain a steady course towards it, using nothing but that external point of reference. Then I was asked to note the relationship of the aircraft nose to the horizon in the pitching (fore-and-aft) plane, and to keep a constant height by maintaining that interrelationship with the stick. Thus I learnt how to fly straight and level. But it also emphasised from day one that there was no need to lock one's eyes on the compass and the altimeter to get from A to B. A frequent check on the cockpit instruments – yes, but the accent was on flying by external reference, and keeping a lookout for other aircraft.

We progressed quickly through these basic handling exercises, to tackle the more vigorous ones like stalling and spinning; both of them preliminaries to tackling the airfield circuit. Fully stalled, the aircraft drops like a stone, like it or not, and since it is flown relatively close to its stalling speed during the approach and landing, it is more than just a good idea to be able to recognise an impending stall, and for recovery action to be an instantaneous reflex. You might argue perhaps that we could have left it at that, but since a stall, uncorrected, can lead to a spin, we practised spin recoveries too. And before getting our teeth into either, we climbed to a healthy altitude.

Recoveries from the full stall made no great demands, provided you had no objection to your stomach being left behind as the aircraft pitched forward like a horse unseating its rider; spins were the really stimulating things. To enter a practice spin (the aircraft completely stalled, the stick held firmly back in the pit of one's stomach and with hard rudder applied) and to watch the fields beneath gyrate into a patchwork blur; then, on the instruction to recover, to apply positive corrective action and a second or two later to emerge crisply from the spin – suddenly all those dreary lectures on Air Force law and the function of the Naval Army and Airforce Institute (the NAAFI), were seen for what they were, mere incidentals to the real heartbeat of the RAF.

At that stage of our training we had an almost blind faith in our instructors, and this read across to anyone who had gained their 'Wings' (qualified pilot's flying badge). And as for those who flew operational aircraft, one instinctively regarded them as infallible, enemy action apart. It was in that frame of mind that we stood outside the flight office one day as a Halifax four-engined bomber approached to land. What it was doing landing at Sywell is a matter for conjecture, but its probable destination was a maintenance unit tucked away somewhere on the far side of the airfield.

The pilot had already made one approach, apparently to size up the rather limited landing run. Now he was making another. An even steeper one than before, aiming to touch down at the far side of the airfield from where we stood, and heading in our general direction. How fascinating it was to watch this large aircraft landing on that small field. It made one blush to think that we should be finding any difficulty whatever in our little Tiger Moths.

As the Halifax rounded out at the end of its approach, it seemed a beautifully judged affair. However, instead of making an early touchdown, as one had expected, the aircraft continued to float serenely across the airfield a few feet off the ground, eating up a considerable distance before it finally came to earth. Surely the pilot knew what he was doing, because had he shared my growing concern about his ability to stop, by now he would already have abandoned the landing and gone round again. No doubt he was about to astonish us with the power of the Halifax's brakes.

Now only three or four hundred yards separated us from the Halifax, and although the frantic squeal of its brakes was clearly audible, it was bearing down on us like a runaway train. Suddenly, even we beginners realised the pilot was in difficulty. At which stage, there were a few stifled gasps from our little crowd of spectators – not all of them from students – as we smartly deliberated which way we should bolt.

Just then the Halifax commenced a gentle swing to its left and, carving its way through the hedge alongside our hut, finished up in the next field facing the way it had come: having spun round on uprooting a small tree with its port inner engine, the latter now drooping sadly from its mounting, dripping oil, but thankfully not on fire. We students looked at each other, rather numbed by this sudden shattering of images.

Within seconds the crew began to emerge from their crippled machine, chattering excitedly and, once clear of the aircraft, to reach for their cigarettes. No doubt they would be serious and sober minded enough when they appeared before the court of inquiry. But for the moment they were no doubt just glad to be in one piece. As the fire tender and ambulance drew up, someone tugged at my sleeve. It was my instructor suggesting that we might ourselves do a little aviating.

I was not among the first to go solo. With my instructor dashing off on leave on the very day I arrived, and his colleagues covering for him during his absence, I hardly enjoyed continuity of instruction. But my day came and I was put up for a pre-solo check: a task traditionally carried out by the Flight Commander or a senior instructor – someone not involved in one's day-to-day instruction. This time there were no niggles or friendly nudges from the front seat; just an impassive silence while I got on with things. After a circuit, an overshoot and a landing, the checking officer got out, carefully secured his harness, and with a brief reminder that I should not hesitate to go round again if either the approach or touchdown was not to my satisfaction, I was on my own.

The drill-like procedures required to fly a tidy circuit are so absorbing that, first solo or not, I found little time to contemplate the absence of my instructor until I was straight and level on the downwind leg. At which point, as I savoured the empty cockpit in front of me, I burst into song. I have never cared for close supervision.

Nearing the end of the downwind leg, as I busied myself checking my heading, the airfield perspective and so on, my joyous outburst tailed off. The cross-wind leg went well; the engine had not died on me as I throttled back; I had remembered to re-trim, and the rate of descent looked about right. As I completed my turn on to the final approach, the speed was a little high and an aircraft which had landed ahead of me was not quite clear of my landing path. I raised the nose a little to kill the speed, checked the trim, and glancing quickly to either side, confirmed that no one was contesting my approach.

The speed was now right, and as I looked ahead I saw to my relief that the landing path was now clear; things were slotting into place nicely. Not undershooting? No. Going to land comfortably into the field, but with no danger of finishing up in the far hedge? Good.

As I crossed the airfield boundary, I began to think about the landing itself – the bit that really mattered. The ground was now quite close, and I switched my attention from the flight panel and the view directly ahead, to scan the grass ahead and to one side. Then, as the ground seemed to rush up to meet the aircraft, I eased back on the stick; checked the throttle closed, and with the speed decreasing, progressively wound the stick back until, finally, the aircraft flopped gently onto the ground. Not quite on three points, but hardly distinguishable perhaps from where my checking officer waited. A few minutes later, as he climbed aboard, he seemed well pleased with my performance. As for me – I was inwardly elated.

At that stage we were almost at the end of our course, and in the day or so preceding my solo circuit, I had – along with others – experienced a cliff-hanging sense of suspense as I wondered if I would go solo before time ran out. Well, I had made it. And although going solo was in itself no guarantee of being selected for pilot training, conversely it appeared that without having done so, one's chances would be pretty slender. Going back to my billet that evening I had a real song in my heart.

During my leave shortly afterwards, my father, although clearly impressed with my news, was not easily convinced that I had managed a flight on my own. I'm not sure why, since I'd always enjoyed being in control of mechanical things. Perhaps events were moving faster than he'd expected, as I'd been in uniform barely five months and was still well short of my nineteenth birthday. Viewed rationally, I had barely begun the process of becoming a qualified pilot, but that did nothing to lessen the pleasure of parental approval.

It was June 1942, and the Commonwealth Air Training Plan was already well established. Assuming my grading school result was good enough, it meant a fair chance of completing my Wings training in Rhodesia, South Africa or Canada (or even the USA, perhaps): free from enemy interference and the vagaries of our climate. With the bonus of foreign travel, it was a time of eager anticipation.

Waiting at Manchester

After seven enjoyable days at Wellington calling on friends and relations, I went north to Manchester to the Aircrew Cadet Disposal Centre to await my grading school result and posting to the next stage of flying training. Several hundred of us lived under canvas at Heaton Park, the grounds of an eighteenth-century hall at Prestwich. Thankfully, the weather was kind and not at all deserving of Manchester's reputation for rain.

For a week or so the waiting was tolerable; amusing too, at times. Mastering the single-seat sculling boats on the park lake fitted into the latter category, since you could even capsize one as you stepped aboard, they were so delicately balanced. More than one of us had a full-scale baptism before we got the hang of them. Neither was Manchester itself, with its cinemas, bars and stage shows, in any sense a dull city. But an airman's pay was limiting, and given long enough is there any place which doesn't begin to pall?

To give them their due, our officers did show some awareness of the dangers of boredom as the weeks went by. They even exploited what talent we had for entertaining ourselves. Having discovered among us a pianist of near-concert standard, they marched us to the local cinema on several occasions to be entertained by his skills: after a while we got to know the 'Warsaw Concerto' rather well. But come July, after a month at Heaton Park, and with no disrespect to our pianist, it was a relief to get away on embarkation leave.

Although we returned refreshed, the tedium which had set in earlier quickly re-established itself. Lord Trenchard himself, the revered 'grandfather' of the Royal Air Force, came to address us – his medal ribbons consisting almost entirely of the rich colours of various orders for the first two rows. But even that sort of fillip could not bolster our morale indefinitely.

As the weeks of waiting turned to months, our energies sought other outlets. An increasing number of cadets were caught climbing in over the boundary walls after lights out, and a few got on the wrong side of the service police for rowdy behaviour. Things reached a climax when a cadet choked to death in his bed after a heavy binge. Or as the Camp Commandment put it more dramatically at a specially called parade, 'one of your number drowned last night in his own vomit, and it is time you chaps exercised some moderation.' We were shocked by what had happened, of course, but felt for our part that it was time authority took some decisions and sent us off on the next stage of our training.

In the last week of August 1942 – ten and a half weeks after I had first reported to Heaton Park – the entire cadet population massed on the parade ground to be segregated according to grading school results and posting destinations. This procedure went on for an agonising thirty minutes, as new squads were formed, before we were told our fate. Happily, I was to continue training as a pilot, and with my new mates was told to be ready to up sticks for a port of embarkation at short notice.

The following day, the C of E padre addressed those of his flock on the merits of taking communion before going to sea. With the U boat menace on the increase – although he did not refer to it specifically – I thought he made a pretty convincing case, and almost for the first time since confirmation, I attended a communion service.

Sea Voyage

I awoke to a sickening sea-saw motion. We had sailed the evening before from Gourock on the Clyde, and according to the best information our destination was Canada. We were ploughing through a short, choppy sea. Occasionally, as we hit a larger wave, the ship seemed momentarily to stop, as it shuddered and creaked under the impact. Later we were to discover the real power and majesty of the ocean when we met the huge rolling waves of the open Atlantic, but as an introduction it was impressive enough.

This was my first time at sea, discounting some mackerel fishing with the Boy Scouts in Watchet Bay. And as I was still absorbing my new environment, the tannoy summoned us to rise and to get into line; first for the ablutions, and later for breakfast. Our sleeping accommodation was tightly packed – up, down, sideways, you name it – and the chaos which would have resulted had we needed to take to the boats in the middle of the night does not bear thinking about. As for the state of the ablutions, hopelessly overburdened as they were, and entirely lacking in privacy, it was a sad day for the squeamish.

For the first forty-eight hours none of us felt too good. But from there on things improved rapidly. Our appetites were back, and we began to take stock of our convoy – its composition, our general heading and so on. We also registered the fact that although we had an RAF OC Troops on board, the ship itself, including the catering, was run by an entirely American crew. And a well disciplined one too, judging from the times we practised assembling at our lifeboat stations.

To fill the day, we were invited to make ourselves useful; and fairly soon I was settled in a pleasantly undemanding job in the officers' galley, a place where food shortages were unheard of and from where, with the chef's connivance, I was able to smuggle a few delicacies back to my friends. One such occasion was my nineteenth birthday, when on return to our quarters, I pulled (rather hastily) from my bulging battledress four or five tins of celebratory plum pudding; having extracted same from the steam oven only minutes before, I can't imagine how I avoided being scalded.

Two of my friends were helping out in the RAF OC troops' office, and responding perhaps to my sharing of the spoils from the galley; they let me in on something they had come across in the course of *their* duties. According to them, my name was on a confidential list of airmen designated as potential officers. My immediate reaction was to invite them to 'pull the other one'. Either way, of course, assuming they were not having me on, it was no guarantee of how things would evolve, but privately it boosted my morale. And, if it proved nothing else, it certainly justified the notion that the Orderly Room corporal is the best informed man in the unit.

1314454 LAC Jordan R.W.

ABOVE. DH82a Tiger Moth used for pilot grading (*Photo courtesy of Air Portraits*) BELOW LEFT. 17 EFTS RCAF Stanley, Nova Scotia. BELOW RIGHT. 17 EFTS flight line of Finch IIs.

After four days sailing, it almost looked as if we were bound for South Africa rather than Canada. It was much warmer, and our daytime course at least was far closer to south than west. Visibility too was greatly improved and we could now make out the total complement of our convoy and escorts. The former comprised four vessels, each of about twenty thousand tons, and the latter consisted of a battleship, a cruiser and eleven destroyers. My classification of our escorts might not be entirely accurate, but the numbers are correct. If our long wait at Heaton Park had to do with mustering this powerful escort, I now saw that frustrating period in a rather different light. All vessels were American, so in spite of our southerly course, our money was still on a North American destination.

If nothing of great interest was taking place within our vessel – pontoon, solo and bridge schools not withstanding – the naval element of our convoy did us proud by way of external attractions. Our escorts practised regularly with their small armament, and most days the seaplanes carried by the battleship and the cruiser made reconnaissance flights in the vicinity of the convoy. These aircraft were launched by catapult and retrieved from the sea by crane; the landing and pick-up calling for a finely co-ordinated drill on the part of the ship's helmsman, the aircraft pilot and the crane operator. With the aircraft approaching the ship's stern, the helmsman applied hard rubber to create a smooth patch of sea, the pilot promptly — touched down, and the crane operator equally promptly snatched up the aircraft before the sea could re-establish itself. As a spectacle it had all the tension and precision of a circus act.

Once or twice we assembled at our boat stations in response to a real alarm signal, with our smaller escorts doing their bit to raise the pulse rate by dashing about, making short, excited, 'whoop, whoop, whoop' blasts on their sirens, and occasionally dropping depth charges. But, thank God, there was never any tangible sign of the enemy. Well, not if one discounts enemy involvement in the fire which took hold of one of our four merchant ships about a day's sailing from our destination. Conceivably this might have been an act of sabotage, but there seemed to be no evidence of an attack from outside. Some week's later we learnt that the ship – although gutted by the fire – had been towed to port and was being refitted.

Twelve days after leaving the Clyde we disembarked at New York. It was early morning when we first sighted land, and some while before we saw the Statue of Liberty and the Manhattan skyline. Perhaps our snail's pace over those final miles through the early morning mist brought things into view too slowly, but somehow the impact of New York from the harbour was something of an anti-climax. Land itself however was a sweet sight, bearing in mind those lurking U-boats.

We did not break our journey in New York. To our annoyance a train was waiting at the dockside: efficiency, we felt, was being overdone. With the minimum of formalities we were heading north for Canada. New Yorkers for the most part received our V signs and cheery waves with a kind of baffled indifference, which seemed a bit odd, America having entered the war alongside us some nine months earlier. Surely, not every one of those sullen faces reflected a personal grudge against the British. Perhaps the New York baseball team had had a bad night.

As we distanced ourselves from New York and proceeded through New Hampshire and Maine, so the reaction became more friendly. From towns with names like Portsmouth and Biddeford, one almost felt assured of a welcome. Nature itself did not hold back one little bit. She had really put out the flags. The maple trees were ablaze with autumn hues of dazzling richness, and the sheer scale of the pine forests and lakes reflected a generous spirit.

Having accustomed ourselves to this grandeur, it was rather a shock to enter the port of Saint John in New Brunswick, which looked distinctly run down and not a little shabby. Moncton, which we reached shortly afterwards, was our immediate destination, and it too had something of a depressed look about it. Most Canadians, as I was to discover later in my service, make no secret of their reservations about the Eastern Maritimes, as compared with provinces farther west. But I knew nothing of this at the time and, as a guest, I was more than willing to withhold judgement at that stage. After all, the view of of most towns from the railway station is hardly inspiring.

Training in Canada

There was no slackening of pace at the aircrew receiving centre at Moncton. We were there just long enough to be split into alphabetical groups – by which arbitrary means our next destination was determined – and to draw some pay in Canadian dollars. The unit belonged to the Royal Canadian Air Force (RCAF), and the atmosphere was brisk and efficient; more consciously so than at an RAF unit, where such blatant keenness would have been considered brash. In short, the RCAF was different. And why not?

Back on the railway, we headed south-west into Nova Scotia, making for No 17 Elementary Flying Training School (EFTS) at Stanley. For the fifty or sixty airmen in my group, whose names began with the letters H to J, there was to be no great trek west to a training school at the foot of the Rocky Mountains, or south-west to Texas deep into the American heartland. We had arrived in Eastern Canada, and that was where we were to remain. The one compensation – we would be back in the air that much quicker.

Once again, as we sped through the countryside, familiar names beckoned – Oxford, Truro and Windsor, and although now and then a name of French origin would remind us that we British had not been alone in discovering Canada, the solid impression was one of English and Scottish settlements. One could hardly have felt homesick in those surroundings, and in that sense it was almost as though the Atlantic did not exist.

If the names were familiar, the trains were not. The engines were huge macho creations which gave warning of their approach over long distances with the wailing of a siren, and in built-up areas with the brisk clanging of a bell. Even at rest those hissing monsters exerted a presence.

A good deal of the journey was spent wrestling with the Canadian newspapers: huge, thick tomes padded out with advertisements and lengthy reports on baseball. News of the war was there, but did not figure in quite the crucial way it was served up at home. It was a graphic reminder that in spite of the familiar place names, we were indeed on a different continent from front-line Britain.

At Stanley, we found the airfield on a slight rise in flatish agricultural land, the latter punctuated with plantations of fir trees. It too was an RCAF station, compact and neatly laid out, without a scrap of litter in sight. In spite of the severely practical lines of the barrack huts and other buildings, the rural surroundings dominated the scene, and being a countryman man at heart, I took to it immediately.

We had arrived over a weekend, and since our first parade was not until Monday morning, we busied ourselves settling into our barrack block. The climate was still fairly mild, but the size of the central heating units, the double windows and the presence of inner and outer entrance doors, confirmed that the winter would be a good deal harsher than the ones we were used to.

The order of priorities on the Monday morning could not have suited us better. Following an address of welcome from the chief instructor, we went almost immediately to the flight line, most of us having a flight later that day.

The chief instructor's welcoming remarks had been far from encouraging, however. Instead of accenting the positive, the burden of his address appeared to be devoted to preparing us for possible failure through lack of aptitude. To be fair to him, parts of his message possibly did not get through because his first language was clearly French rather than English, but his address was peppered with the words 'cease train', and I'm afraid that is how he was known from there on. Far from providing inspiration, it was several days before we had shaken off the gloomy effects of his 'welcome.'

My personal instructor, on the other hand, was a breath of fresh air. He was a man of decision, not given to over-long explanations, and certainly not given to pessimism. We clicked. He flew with a style in keeping with his personality, a style which managed to combine crispness with fluency, but which stopped short of anything flashy. He was also quick to sense when it was time to impose some additional stretch on my budding talent, so that flying with him was a pleasure and sorties rarely dragged.

The aircraft were Fleet Finch radial-engined biplanes. And although they were more up-to-date than the Tiger Moth in most respects, the Finch's engine was hardly in the van of modernity. Its cylinders stuck out proud of the cowling, and once hot, they habitually threw oil onto the windscreen; the globules getting darker and bigger as the sortie progressed. So that in the last fifteen minutes of a one-hour sortie, the windscreen was often liberally spattered. But it never really impaired forward vision, and, once we had learnt to live with it, it even added a touch of romance as our minds went back to the oil-soaked fighter aces of the First World War.

The flying programme was maintained at a brisk tempo. We often flew four sorties in a day, and occasionally, when sorties were lost through bad weather and the programme as a whole looked like slipping, we made up the loss over the weekend. Going solo on the new type, although not regarded with quite the same awe as one's solo at the Grading School, was still something of a watershed and cause for a modest celebration.

Once we had mastered the airfield circuit, and with our solo flight safely behind us, we set about more ambitious manoeuvres. The allocation of twelve flying hours at our Grading Schools had been too precious for any of it to be frittered away on such things, but at Stanley, with between sixty and seventy hours to play with, the task was to develop us more completely as pilots.

We entered, for the first time, the glamorous world of aerobatics, where one minute we hung upside down in our harnesses in a slow roll – the blood swelling the veins in our necks and bits of muck from the cockpit floor falling onto the inverted canopy – and the next we were at the top of a loop, being held firmly in our seats by a strange force referred to as 'g'. Even the boyhood thrill of those huge fairground swings paled as we began to explore that wonderfully unfettered world.

To begin with, some of the manoeuvres were not at all easy to perform with precision, but we did our best, and even if we went off line in a loop or came out of the final stages of a slow roll with a bit of a 'whoosh' and with the nose below the horizon, we were still expressing ourselves with a new and exhilarating freedom.

Naturally enough, there were bouts of frustration, as we hit a plateau in our progress. But the overriding feeling was one of fulfilment. Solo flights in themselves induced a wonderful feeling of liberation, and at the end of a good day's flying the sense of well-being lingered on like a mild form of intoxication.

As part of the ground syllabus, we were introduced to the Link trainer, a classroom device for teaching instrument flying and the use of radio aids. It was housed in its own special room, kept as dust free as possible to protect the machine's delicate mechanism. The linoleum floor was polished like a sheet of glass, almost as though the threat of a high-powered inspection hung constantly over the place, and we students were expected to play our part in maintaining this purity. Entering the room, we changed into plimsoles, with the instructor going one better by skating about on felt pads, such was the accent on gleaming cleanliness.

The Link student sat in a replica of an aircraft cockpit, free – within limits – to yaw, pitch and roll. Seated in the cockpit, the student lowered a hood to cut off his external view, and from there on flew the Link solely by reference to the flight instruments and radio signals fed to him through his earphones. Meantime, the track of the flight was traced in ink by a crab-like instrument on a table a few feet away. The flight over and the hood raised, one dismounted to view one's efforts and to receive a critical analysis from the instructor. How very obvious our gaffs were, viewing them on the table, as compared with minutes before inside the confusion of that wretched box.

There were moments when most of us regarded the Link trainer as little short of an instrument of torture. On such occasions, it seemed that the harder one worked to maintain the correct instrument readings, the more likely you were to over-correct with the controls, and to end up chasing the instrument needles in a frenzy of activity. Meanwhile, the machine itself would add to the general state of agitation by hissing angrily at frequent intervals, as its control bellows were made to work overtime. The fact was that the Link did not respond exactly like an aircraft, and to control it really well required a knack. Once that was accepted, the torture could at least be reduced to a tolerable level.

Being close to the Atlantic seaboard and possessing a 'metalled' runway, Stanley was visited occasionally by operational aircraft. Whether such visits always resulted from official operational policy, or were sometimes the result of someone dropping in on an old friend for lunch, I'm not sure. But such visits invariably created considerable interest among staff and cadets alike. Even so our visitor was usually something fairly prosaic such as a Lockheed Hudson engaged in maritime patrol duties. So on the day we receive a flight of Hurricanes, the flight line fairly buzzed with excitement.

Everyone dropped what they were doing, to watch the Hurricanes taxi in and to see how operational fighter pilots handled their aircraft. It was all very impressive until one of them put a wheel off the taxi track and became bogged down. The aircraft was not stuck for long, but what should have ended as a slick demonstration of professionalism had unfortunately been marred, and my mind went back to the anti-climax we cadets had experienced after the Halifax incident at Sywell, not that the consequences to the Hurricane and its pilot were in any sense comparable. It may be unreasonable of young men to expect that their superiors should never put a foot wrong, but we had still to grasp that point, and most of us were only too ready to see it as a case of 'finger trouble'.

In the last two weeks of the course it gave a great infusion of interest to set off on a series of cross-country flights; how grown up we felt escaping from the mothering attentions of the local flying area. Practice in maintaining an accurate heading and the correct use of the map were important objectives of these exercises, but the overriding lesson to be absorbed was the need for thorough pre-flight preparation. Before we set foot anywhere near the aircraft, we had already taken the map and pencilled a ring around the prominent features along our route, noting alongside the estimated travelling time. So as each feature became due according to the watch, we knew precisely what to look for. The contrary procedure of seeing a feature on the ground, then trying to locate it on the map, was looked upon as a sure way of getting lost and nothing less than an invention of the devil.

It was during these excursions that one first sensed the limitations of the Fleet Finch as a means of travel. Its cruising speed was a mere eighty-five miles an hour, so you only needed a fifteen mph head wind to be overtaken by a fast car. But there was certainly one advantage to this leisurely progress: combined with the fact that the towns were well scattered, it meant that the risk of mistaking one place for another hardly existed.

To begin with, these cross-country flights were instructional sorties, during which I was kept busy answering my instructor's questions about track error, arrival time at the next turning point, and so on. Later, cast loose on my own, and with an airfield no longer close to hand, I was suddenly aware of how much I depended on that oil-spattering single engine. True, I was too busy applying the navigation methods I had lately been shown for this awareness to become anything like an obsession. But at odd moments over the wilder stretches of the route I confess to listening rather intently to the engine's beat, as well as to keeping half an eye open for possible forced landing areas. It has to be said, the pine forests looked most inhospitable.

At about this time we also had our first taste of night flying, which apart from the general transformation brought about by the darkness, gave us plenty of new things to think about in terms of the airfield lighting, our greater dependence on the flight instrument panel and the tighter control procedures. But, on the credit side, I was also struck by the beautifully smooth conditions once the sun had gone down.

My main preoccupation, however, was how my defective colour vision would stand up. The answer in a nutshell – well enough. I was relieved to find I had no problem distinguishing between a red and a green signal from the runway controller's lamp at the take-off point. However, I was much less confident about telling amber from red on the angle-of-approach indicators coming in to land. But this did not defeat me. I could recognise the green segment well enough – which denoted the correct approach path – and if I was so high as to be in the amber segment or, alternatively, so low as to be in the red segment, a glance at the altimeter and the perspective of the runway lights, told me which of these situations I faced.

I also had some difficulty with the colour of the Very pistol lights occasionally fired from the runway controller's caravan. But almost invariably other circumstances were present to suggest whether it was a red – usually the case – or a green. And if I was in doubt, I opted for safety. In sum, my colour deficiency was a disadvantage but not a danger, provided I kept my wits about me. Needless to say, I did not discuss the matter, even with my closest friends.

There was no measured pause between our elementary training and the next stage. We made the journey to No 8 Service Flying Training School (SFTS) at Lakeburn in New Brunswick over a weekend, and resumed training promptly on Monday morning. Our new toys were twin-engined Avro Ansons, Canadian-built and powered by American-manufactured Jacobs engines.

After the compactness and simplicity of the Finch, the Anson looked slightly intimidating, what with the extra instruments, the additional controls for the second engine, the more complex fuel system, the flaps and the retractable undercarriage – not to mention its sheer bulk. But we were soon at home in that docile, forgiving creature. And after a while, I suppose our only real complaint was that it was almost too docile; whereby its wings shed lift so reluctantly during the pre-landing hold-off, that a clean three-point touchdown was almost impossible to guarantee.

The Anson was a fresh challenge, and welcome for that. But I did feel some disappointment at not going on to the single-engined Harvard, which would have put me in line for operations on fighters. As it was, Anson training would almost certainly mean a posting to a bomber or a maritime operational training unit, which I saw as much less exciting, and perhaps a reflection on one's ability to fly an aircraft to its limits. With hindsight, it seems clear that our EFTS course was sent almost en bloc to a twin-engine training unit as part of a general switch to offensive operations. The die cast, I did not nurse my disappointment.

While at Stanley our course had consisted entirely of RAF cadets, at Lakeburn RCAF and RAF cadets were pitched together. This integration extended to every aspect of life, from sharing a barrack room to sharing flying instructors; and, once past the solo stage, flying together alternately as pilot and crewman on navigation and bombing exercises. As to the differences in national make-up, I found little of importance separating us, and instinctively I took to the easy-going and uncomplicated ways of the Canadians. How they viewed our more guarded approach and liking for understatement, I'm not sure. But reservations quickly disappeared as we accepted one another as individuals.

Of course some basic characteristics remained to distinguish the two communities. In fact, these very differences added some welcome spice to the scene, and we would have been the poorer without them. The colourful flow of language used by the Canadians, for instance, when something went badly wrong. Not for them our use of a single sharp expletive. They gave vent to a whole string of words, which seemed to hang in the air and which left no doubt as to their frustration. You will admit, I think, that phrases like 'God damned son of a bitch' have a certain backwoods quality, rarely encountered in an English town.

To return to the matter of flying, my concept of how one might handle the throttles when turning a twin-engine aircraft in the air was quickly dispelled. Until then, I had vaguely imagined that one might open up the engine on the outside of the turn, to help the aircraft change direction, just as twin-engined aircraft are manoeuvred on the ground. Flying Officer Yunker eyed me suspiciously when I asked him if that was how it was done, as we walked to the aircraft for my first flight. I suppose it must have crossed his mind that I was pulling his leg, but after a moment's hesitation he explained the position. The throttles are not handled in that way of course, the force required to turn the aircraft being initiated by applying bank and following up with enough rudder to prevent any sideslip. After that early loss of face, I kept my theories to myself.

The Anson was a good deal less fun than the Finch. Aerobatics, spinning and practice forced landings, gave way to serious stuff like additional instrument flying, extended navigation exercises and simulated engine failure. Things hardly designed to bring the colour to your cheeks, but necessary if we were to apply our skills in some practical way. Towards the end of the course, practice bombing and night flying relieved the routine. But on the whole, the Anson course demanded diligent application rather than flair, the sustaining factor being the promise of our Wings.

We had arrived at Lakeburn in early November and were soon into winter, with night-time temperatures down to thirty degrees below zero Fahrenheit. Overnight, the aircraft were housed in heated hangers, and come the morning, as each was pushed into the open, its engines were started within seconds. Reaching the flight line, those engines were kept running until the pilot took over for the first flight. And with subsequent between-flight delays cut to a minimum, the engines were never less than warm throughout the entire working period. At the end of the day, the final dodge – to assist start-up in the morning – was to dilute the engine oil with petrol before shutting down.

Aircraft operation was not the only thing threatened by those extremely low temperatures, as a member of our course soon discovered. Marching briefly from our barrack hut to the lecture hall in a biting wind, he chose to ignore the advice to lower the earflaps of his RCAF Yukon cap, with painful results. That evening, one of his ears was half as big again with frostbite. He was dating one of the station WAAFs at the time and had wanted to keep up appearances, but even he had to admit that his vanity had backfired on him on that occasion. Fortunately he suffered no lasting injury, but everyone took the point that if you waited until you felt the need to cover ears and hands, it might already be too late.

Once the winter snowfall was into its stride, it accumulated so rapidly that no further attempt was made to clear the runway until the spring thaw. Landing on a foot or so of compacted snow worked like dream as long as you stayed on the runway. If you veered off, too bad. Waiting menacingly on either side was a solid bank of frozen snow; cast there earlier by the snow plough. A less obvious danger was the insidious encouragement – given by the slippery surface – not to worry if you landed with sideways drift. On a runway free of ice and snow, such sloppiness would have caused many an anxious moment, and a few damaged wing tips; a fact confirmed with the thaw, when some of our course were slow to adapt.

To achieve our quota of flying hours called for constant endeavour. Never more so than when a spell of bad weather suddenly lifted and there was a mad rush to get into the air. I recall one such instance with some amusement, although it was not at all funny at the time. It came at the end of an idle morning, just as I had finished yet another bottle of Coke. Every available student was suddenly despatched solo to cram in an hour or two while the weather held. Twenty minutes later I was bursting to relieve myself, with no one to take control whilst I did so. It was then that I discovered the knack of trimming the aircraft just enough nose down to compensate for my visit to the facility at the rear of the cabin. A knack not achieved without some agitated trial and error. The object achieved, the feeling of relief was unbelievable.

There was one exercise in which even the stately Anson shed its corsets – low flying. The flat, thinly populated countryside of New Brunswick was ideal for this pleasant pursuit. And the onset of winter, with everything clothed in a mantle of white, heightened that pleasure as we skimmed low over frozen lakes and swept past pine forests glistening with snow. Now and then we might spot a herd of deer. But generally the scene was one of peaceful hibernation. It seemed to me that my instructor, given his staunch observance of the minimum height regulations, did not enter fully into the spirit of the occasion but, despite his inhibitions, those flights left me thoroughly invigorated.

We were paid fortnightly, and for most of us pay day meant some minor extravagance, like a beer or two in the canteen, or a trip to Moncton for a meal, followed by a visit to one of the cinemas or dance halls. But for a number of the Canadian cadets – much better paid than ourselves it's true – this was altogether too tame.

Immediately after tea, each clutching a fistful of dollars, upwards of half a dozen of them would gather in the ironing room to stake their pay on a game of 'craps'. Preparations were minimal. The ironing table was pushed against the wall, it surface draped with a blanket. The bets placed, a pair of dice was shot noiselessly onto the table to rebound off the wall. But to backtrack for a moment. Before the true aficionado released the dice, there was a piece of ritual to observe. As he shook them vigorously in his cupped hands, he 'spoke' to them, announcing the number he wanted: these preliminary incantations – consisting largely of rhyming slang – being delivered with increasing fervour as the game developed.

The first throw made, sometimes money changed hands immediately; but mostly, the thrower would have to repeat the number without first coming up with a losing combination. Since losses meant doubling up in order to recoup, the atmosphere became more and more electric, with money changing hands at a furious pace in the closing stages. From time to time the dice changed hands until, finally, one person had virtually cornered every dollar in the room; the whole process being over in some ninety minutes. With the majority of players emerging to face the next fortnight hard up or on borrowed money, it was not easy to see why the game had such a lasting appeal. But you could be certain of one thing – the same people would be back in the ironing room a fortnight later, confident that this time the dice would roll their way.

As an occasional spectator at these proceedings, it struck me as a crazy way of running one's finances, but later in my wartime career, with cash to spare, I again fell in with some Canadians and discovered for myself the pulsating excitement – as well as the pain – of playing 'craps'.

Christmas in Montreal

As we approached the Christmas and New Year holiday, we had not had a decent break since beginning our elementary training in mid-September. So when Charles Medland – a fellow student – invited me to join him for Christmas with his relatives in Montreal, I jumped at the chance. For two weeks Bing Crosby – courtesy of the Bulova Watch Company – had all but monopolised the air waves with *I'm Dreaming of a White Christmas*, and I was nicely conditioned for a family occasion.

Memories of the train journey have faded with time, but I recall that it stretched over many hours and that as the journey progressed we became increasingly glad of our greatcoats. My strongest recollections, however, concern one of our travelling companions – another airman.

Airmen as a breed are hardly reticent, but this particular one was garrulous in the extreme. We had not travelled a hundred miles before his boastfulness was becoming quite a pain. Part of his problem appeared to be the presence of a girl, whom he was obviously out to impress. He must have realised that his stories were getting increasingly improbable, but by then he was apparently unable to stop himself. Finally he looked to box himself in, and there seemed no doubt that shortly his bluff would be called.

At that time you could not purchase liquor in Quebec Province except at a state shop, and there were strict formalities to be observed. 'But such stupid laws need not bother us,' said our chatty friend, 'at the next stop I can get whatever drink you want, without any formalities.' Charles and I exchanged winks and said a whisky would suit us nicely.

Shortly afterwards (at Mont Joli, I think it was) it was he and not us who had put one across. During our thirty-minute halt, he led us unerringly to a source of bootleg liquor – the local undertaker. Surrounded by coffins draped with dust sheets, and joined by the undertaker himself, we sipped fiery whisky from china mugs, half expecting the law to walk in at the moment. My other concern was that we might inadvertently catch sight of a corpse before the party was over. But in the event, neither of my fears materialised. As we recommenced our journey, we forgot the lack of heating and saw our companion in a kinder light.

Our reception in Montreal was warm and generous, and there was no letting up in the display of hospitality. We stayed with Charles's uncle, who started things rolling with a big party. As the days went by, it seemed that every relative and close friend present had laid claim to a return match, and in the several days we were there, there were as many parties. Without in any sense wishing to appear ungrateful, I think that secretly, our hosts must have been as relieved as we were when it was all over, it was such a marathon. The fact was that every one of them wanted to be identified with the 'Old Country' – a country which had come through on its own in 1940. For them, during the brief period of our stay, Charles and I represented something they admired. It was quite humbling, and since neither of us had seen action, it also made us feel rather bogus.

The return journey found us drying out and catching up on sleep. Even our garrulous friend appeared subdued. He did, however, rouse us as we approached the stop where the undertaker lived, only to receive a polite but firm refusal.

Wings

On our first working day of 1943, I logged four hours in the air. With our Wings graduation day now only two months away, and refreshed by the break, it was good to be back in the cockpit. Later that week we began to put real flesh on the bones with our first taste of formation flying and practice bombing.

We shared Lakeburn airfield with Trans-Canada Airlines (TCA), so that, at night especially, we had to be alert to the arrival and departure of the scheduled TCA passenger flights. If any one was to give way, it clearly had to be us, emergencies apart. Sharing the airfield with TCA also helped to keep our self-esteem in check as we watched the polish and assurance with which they handled their machines.

February's flying programme was punctuated by our all-important end-of-course tests. One of these was a fatigue test, where one took off on instruments under a canvas hood and, blind to external references, remained at the controls for over three hours. Robbed of the natural horizon, one's instincts sometimes strongly contradict the aircraft attitude displayed on the instrument panel, and the object of this particular test was to check that we could still resist these misleading instincts as fatigue set in. (Sometimes, I felt that the first barrier to be overcome in simulated instrument flying was that claustrophobic hood itself.)

In parallel with the last of our flying tests, we sat our final ground-school examinations in the numerous supporting subjects. It was a relief when February was over.

March was a watershed. First came the Wings graduation ceremony, at which that coveted badge was formally pinned on our tunics. And a week later my commission came through. Once both were in the bag, the euphoria was complete.

My ambition now was to get home and on to operations. The fact that I was considered bright enough for a place on the next General Reconnaissance (GR) course on nearby Prince Edward island – whilst flattering – did not interest me at all. My instructor did his best to persuade me of the advantages (more flying experience before assuming operational duties, etc), but since GR graduates inevitably finished up in Coastal Command, and I could think of nothing more dreary than patrolling umpteen square miles of featureless ocean, for me it remained a non-starter. I'm pleased to say that no attempt was made to coerce me.

For those of us who were commissioned, the narrow braid of a pilot officer no longer looked quite as inconspicuous as we sized ourselves up in the fitting-room mirror. And being commissioned made an extraordinary difference to our everyday lives; a difference which I had not given thought to as we worked towards our Wings. Suddenly one was being treated as an individual, and had a style of living to match. It would also make extra demands, of course, but for the moment we had not taken on those extra responsibilities and the pleasurable aspects of status were undiluted.

Only a week before our graduation ceremony, I had completed several days 'Jankers' for allegedly selecting an undercarriage up on the ground; my punishment being to clean out the station guardroom every evening for a week. Either my standing on the course had been sufficiently secure for my commissioning to survive that incident, or the personnel staff work had been sloppy and had failed to catch up with events. Whatever the reason, the ignominy of that recent punishment made the trappings of my commission even sweeter.

ABOVE. Wings graduation day 8 SFTS, 5 March 1943. BELOW LEFT. Wings and commission euphoria. BELOW RIGHT. Bill and Harry with 'Mom' and 'Pop' McKenzie.

ABOVE. Wellington 1c *(Photo courtesy of Imperial War Museum)* BELOW
LEFT. Front page of our leaflets. BELOW RIGHT. OTU escape photograph.

LE COURRIER DE L'AIR

APPORTE PAR LA R.A.F.   LONDRES, LE 21 OCTOBRE 1943

Rupture du front allemand sur le Bas Dnieper

LA LIGNE DU DNIEPER, DÉSESPÉRÉMENT DÉFENDUE PAR LES ALLEMANDS, EST EN TRAIN DE S'EFFRITER SOUS LES COUPS QUE LUI PORTENT LES RUSSES ENTRE GOMEL ET ZAPOROJE.

Dans une zone au moins des têtes-de-pont russes sur la rive occidentale du fleuve — celle de Krementchoug — la bataille a dépassé le stade de la tête-de-pont, les Russes ayant rompu les défenses allemandes sur un front de 50 kilomètres et avancé d'au moins 45 kilomètres en direction du sud-ouest.

Au cours de cette avance, qui se poursuit, ils ont occupé l'embranchement ferroviaire de Pyatikhatka, où l'une des lignes d'évacuation de Dnepropetrovsk bifurque vers Znamenka et Krivoï Rog.

Au sud de Gomel, les Russes ont forcé le passage du Dnieper en aval de son confluent avec le Soj, sur un front de 20 kilomètres, et progressé au-delà du fleuve de 3 à 10 kilomètres. Il semble qu'ici les Russes aient l'intention de remonter la rive droite du Dnieper, vers Rechitsa, à 40 kilomètres à l'ouest de Gomel. L'occupation de Rechitsa, sur la ligne de chemin de fer de Gomel à Pinsk, rendrait presque intenable l'évacuation de Gomel.

Dans le secteur de la tête de pont au nord de Kieff, les Allemands lancent des contre-attaques importantes, qui jusqu'ici ont été toutes repoussées. Ces attaques allemandes sont significatives. Elles ressemblent fortement à celles que l'ennemi déclencha pour tenter la prise de Kharkoff par les Russes. À Kharkoff, elles échouèrent avec des pertes sanglantes, échec dont les conséquences directe fut la chute de Kharkoff. L'expérience montre...

UNE VUE DU CHAMP DE BATAILLE DE GOMEL

Photographie prise après la rupture des lignes allemandes par les Russes. À l'arrière plan coule le Sozh.

EN ITALIE LES ALLIES AVANCENT

Les Allemands ont abandonné toutes leurs positions fortifiées de la ligne du Volturne.

La Ve Armée progresse dans la direction du Garigliano. La ville d'Isernia, située sur la route de Rome, est menacée par les Alliés.

Après de violents combats, la VIIIe Armée a capturé Campobasso et Vinchiaturo. Elle a désormais la maîtrise de la route importante Termoli-Vinchiaturo.

La Ve Armée déclencha son attaque de nuit le 12 au 13 octobre. Le passage du Volturne, rendu difficile par la crue, fut effectué en trois points différents.

Le lendemain et les jours suivants, les positions, faisant donner toutes ses mitrailleuses et tous ses mortiers, et livra bataille avec acharnement sur la plage de Salerne. Mais, sous la protection d'un puissant barrage d'artillerie, les troupes alliées traversèrent le fleuve, large de 50 mètres environ, avec leurs chars d'assaut et leurs tanks.

Pendant ce temps, quelques éléments de la Ve Armée débarquaient de la côte, au nord de l'embouchure du Volturne, tandis que les contre-torpilleurs de la Royal Navy qui les avaient amenés à quai d'ivoire, canonnaient les positions ennemies pour couvrir le débarquement.

Le lendemain et les jours suivants, l'ennemi, résistant toujours âprement, reculait sur toute la ligne de la côte ouest aux Apennins. Cependant, les avia-

...tions alliées attaquaient le système ferroviaire de l'Italie du nord et du centre ainsi que les transports contenus à l'arrière du front.

Les lignes de communication allemandes entre Rome et le Volturne furent disloquées par des bombardiers de l'Aviation tactique à Sparanise, Valvaira et Venapo.

En avant du front de la Ville Armée, des Spitfires et des Kittybombers de la R.A.F. et de l'aviation australienne, ainsi que des Warhawks américains, effectuèrent plusieurs raids sur la ligne de chemin de fer côtière entre Pescara et Ancône.

D'autre part, la Luftwaffe s'efforça, mais en vain, de démolir les ponts jetés sur le Volturne par le Génie de la Ve Armée. Une quinzaine de bombardiers-chasseurs concentrèrent leurs attaques sur un pont sans parvenir à l'atteindre. Sept et même trois d'entre eux furent abattus par les canons de la D.C.A.

Il semble que les Allemands vont essayer d'établir une nouvelle ligne de résistance sur le Trigno. Quoi qu'il en soit, les armées alliées possèdent maintenant des positions importantes qui leur permettront d'avancer sur Isernia le moment venu.

La victoire des Patriotes corses

Le général Devanque, Chef d'État-Major du général Giraud, a récemment fait à Alger un exposé donnant des précisions sur la Bataille de la Corse.

En trois semaines de combat, le territoire du premier département français, comptant 280.000 habitants, a été libéré. Voici comment cette libération fut effectuée.

Les troupes régulières ont agi en action étroite avec les patriotes, qui s'étaient organisés spontanément puis minutieusement groupés depuis le printemps dernier sous la direction de Colonna d'Istria. L'armée de terre et la marine ont pris une part effective au combat, ainsi qu'un groupe de commandos américains. Ajaccio, située à 160 kilomètres d'Alger, fut tout d'abord libérée par les patriotes pour permettre

(Suite à la page 2)

La Conférence de Moscou

À Moscou, MM. Eden, M. Cordell Hull et M. Molotoff ont commencé leurs travaux le 19 octobre.

À cette conférence la stratégie politique alliée sera discutée ; en raison du vaste champ de ces discussions et il est probable que la conférence sera prolongée.

Les délégations ont fait savoir qu'aucune information ne serait donnée, soit par la radio, soit par la presse, avant la fin de la conférence. Si des décisions prises sur la stratégie politique le même temps qui s'impose à propos des décisions dans le domaine de la stratégie militaire.

When the NCO in charge of the guardroom first greeted me in my new capacity, neither of us could suppress a smile at my sudden change of fortune.

The only anti-climax of that period was the long wait for a ship to take us home; shades of Heaton Park. But for a kindly local couple – 'Mom' and 'Pop' McKenzie – who gave open house to three of us (Bill Hearn, 'Harry' Harrington and myself), it would have been a very dull time; as one, then two, and finally three weeks went by. The warmth of their hospitality was a re-run of my Christmas break in Montreal – less the accent on alcohol, thank the Lord. Their show of hospitality was all the more touching since 'Pop' was retired and of modest means. Such was the extraordinary regard for the 'Old Country'.

Our return to Britain contrasted sharply with the twelve-day outbound journey. On that occasion our safety had depended on the strength and vigilance of a large escort and, no doubt, on a degree of circuitous routing. Now we relied on sheer speed. The *Queen Elizabeth* took us, *unescorted*, from Halifax to the Clyde in four and a half days; the intrusive throb of her engines never absent night or day, as she made her dash at near maximum power. Our lack of escorts was hardly a matter of bravado – after all, what could they have found to shepherd a vessel capable of thirty knots?

We disembarked on 4th April and a day or so later I knocked on the door at home and walked in; my arrival otherwise unannounced. My parents were a bit bowled over by my sudden appearance. It was breakfast time, and my father had to leave for work after only a brief exchange. There was not even time to explain that since my cable giving news of my Wings, I had also been commissioned.

During the morning, the Great Western Railway delivered my heavy trunk, and with some help from my three-year-old sister Sheila, I unpacked the perishable items there and then in the kitchen. Gasps of delight and approval punctuated this process, as I produced one handful after another of milk chocolate bars and a jumbo block of processed cheese, wartime rationing accounting for this reaction to mere groceries and confectionery.

My father returned promptly for lunch, and asked me to settle the question of my rank. Evidently one of his friends – who happened to see me arrive at the railway station – had contradicted him when he referred to me as a sergeant pilot. The situation explained, Father's face beamed with satisfaction. But there was also a hint of disbelief as he absorbed the news. Since I had failed the *scholarship* examination and left school at fourteen, I suppose this sudden finding of form did require some digesting.

For a few weeks following my disembarkation leave, my parent unit was a Personnel Reception Centre in Harrogate, that stylish Yorkshire town with a nice nip in the air. Upwards of two hundred and fifty graduates of the Commonwealth Air Training Plan were billeted in the Queens Hotel, while the air staffs and administrators consulted over what should be done with us. Sympathising with their dilemma, most of us contrived to work in a little additional leave, so as not to add to their burden. But by the time the paper shuffling was over and we had dispersed to tackle the next stage of training we were more than ready for some honest work.

Consolidation Training

From May to August I built up another hundred flying hours as I consolidated my skills alongside other newly qualified pilots.

The main object was to polish up our night flying and to give us practical experience in the use of radio beams for airfield approach purposes, something we had so far only practised in the pseudo atmosphere of the Link trainer. But we soon discovered that flying in UK airspace was a challenge in itself. Gone were the wide open spaces by day and the brightly lit towns by night. What we now faced was a densely populated island littered with high ground, barrage balloons and dummy airfields, and subject too to a rigorous blackout. Add to that the vagaries of the English weather and the sometimes too-well-camouflaged state of one's own airfield, and you can see there was plenty to stimulate us johnny-come-latelies.

My first commitment, however – a refresher course on Tiger Moths at Sywell – was more in the nature of a gap filler, but an enjoyable gap filler for all that. Being billeted with a sociable family meant that the evenings passed quickly, and by day I had the good fortune of being schooled by one of my former grading school instructors. The latter was wonderfully indulgent of my wish to experiment with the limits of manoeuvre, although having done his best to oblige with an inverted spin – of which the Tiger, incidently, was having none – he did finally declare that there were perhaps other, more practical, aspects of my technique which we should be working on.

My Tiger Moth course was not without incident. In addition to us newly qualified pilots, Sywell had also acquired an intake of fledgeling navigators, the idea being that after the pilots had reacquainted themselves with the Tiger Moth, they would pair off with the navigators to do some combined training in cross-country navigation. As a practical solution it could hardly be faulted. And Lord knows my navigator, Pilot Officer Bobby Bannister, seemed as bright as they come.

On a sunny day with less than half cloud-cover, we set out on a simple triangular route. With the luxury of a navigator on a flight I would normally do on my own, this was going to be a real doddle. Therein lay the trap perhaps.

An hour after take-off I was obliged to land at an airfield to ask our whereabouts: wartime airfields, like wartime railway stations, having been denuded of name boards and the like. We had fetched up at Staverton in Gloucestershire, miles off track. And although we got back to base without further difficulty, it was only by sticking firmly to the principles of pilot navigation. By then I had dropped any sense of deference to Bobby's specialist qualification and had my own eye firmly on the ball. Having to explain what we'd been up to, once back at Sywell, also taught me that navigator on board or not, it's the aircraft captain who carries the can.

For the remainder of my consolidation training, I reverted to twin-engined flying, for which I was based principally at Little Rissington in Gloucestershire. Our machines were Airspeed Oxfords; lively, likeable aircraft, with terrier-like qualities; less forgiving than the Anson and that bit more challenging if one was to fly them really well.

One of the more outlandish training techniques inflicted on us during this phase was that of 'Day/Night Flying', whereby we attempted to practise night-flying techniques in broad daylight. With our instructors sitting alongside us enjoying the sunshine, we students donned a pair of dark goggles and practised landing on a flarepath consisting of brilliant yellow lights. Apart from the fact that the goggles frequently misted up (in spite of a ventilation tube which we stuck out of the cockpit window), the whole business felt most peculiar. And since we afterwards completed a course dedicated to night flying as such, I'm not sure what purpose those weird night-simulation sorties were meant to serve, but no doubt the boffins gained some satisfaction from poring over the results.

Little Rissington also gave me my first taste of a permanent RAF station. With its handsome tree-lined roadways, its superb sports facilities and palatial officers' mess, I was slightly in awe of the place to begin with. But what a change officers' mess life represented with its pleasantly cosseted atmosphere: sharing a bedroom with only one other person; drinks brought to the ante-room by a punctillious white-coated steward; being waited on at table; a billiard room with three full-sized tables. Coming within weeks of two-tiered bunk life in a barrack room, and queuing for practically everything, such amenities appealed to me more than I care to admit.

Operational Training

In early September, I reported to 15 Operational Training Unit (OTU) at RAF Harwell; a bomber command unit whose role was to feed the Halifax and Lancaster squadrons with trained crews.

By then, the bomber command offensive against Germany was already well established – the first raid employing a thousand aircraft having taken place over a year before. However, with the powerful German air defences (radar, searchlights, flak belts and night fighters) well organised and in depth, our bomber crews were having no picnic. Casualty rates were frequently between thirty and forty aircraft per raid, and occasionally between sixty and ninety. Translated into aircrew losses, these amounted to some two hundred to four hundred and fifty personnel per raid. It was against this background that 15 OTU, with its fleet of cast-off Wellington bombers, performed its role.

The job facing me was not merely to learn to fly the Wellington, but together with the members of my crew, to make use of it in learning the rudiments of the bomber role. That is: to get safely into the air with a full bomb load; to navigate to the target; to deliver the bombs accurately and on time; to shake off the enemy fighters; and to return safely, so as to repeat the job a night or two later. Initially, this involved each of us in a period of separate ground training, during which – other than getting an overview of the aircraft – we made ourselves thoroughly conversant with that equipment peculiar to our individual function.

Every one of our instructors, ground and flying, were men who had survived thirty raids or more into the German heartland. They had regularly watched those less lucky than themselves spiral down in flames and had grown used to the empty places at the breakfast table the next day. And it had left its mark. Now they were dedicated to passing on their hard-won experience and instilling in us that degree of professionalism we too would need – luck apart – if we were to survive that hostile environment.

Among the ground instructors, I particularly recall Flight Lieutenant Snowden, who showed extraordinary commitment to dinning into us the aircraft systems, cockpit checks and emergency drills. Before he had finished with us we could have drawn the aircraft systems blindfold and recited the emergency drills in our sleep. As to Snowden the person, he was extremely tense, and seemed almost afraid to show warmth or humour. Perhaps that was partly his nature, but to us it looked more the result of nervous strain. It was clear that before long, more than our flying ability was going to be tested.

We were encouraged to form ourselves into crews by mutual selection. The only stipulation being that this process should be completed by the end of the two-week ground school period. For my part, I had a crew well within the first week, which brought some adverse comment from an older colleague, who thought I had been altogether too hasty in my judgement. I felt a bit wounded by his remarks, until he revealed that he too had had his eye on my navigator, when I moved in and snapped him up. In this case, my choice proved to be a good one, but one of my selections was to prove a big disappointment; of which, more later.

The ground school period behind us, a number of us moved to the satellite airfield at Hampstead Norris, where – as with practically any outstation – the atmosphere in the Mess was altogether less stuffy, and the domestic routine of the station noticeably more relaxed. Administratively, Hampstead Norris was run by a camp commandant, who seemed to have just discovered the tannoy system and, as with a new toy, was unable to leave the thing alone. Unwittingly he gave us a great deal of amusement, as every half an hour or so, we were treated to his latest broadcast of administrative trivia. In his case he really did live up to the nickname of 'Camp Comedian'.

But these relaxed standards were not allowed to detract from operational activities. In the things which mattered, Hampstead Norris was very much a bomber command station; the aircraft being well dispersed – reflecting the lessons of the Battle of France – and a tight discipline being maintained over the flying programme.

One result of each Wellington having its own dispersal was that the aircraft's lines were not swamped by sheer numbers. It may not be essential to like the look of your new aircraft, but it's not a bad start if you do. I had long admired the Wellington's elegance as it had flown overhead, and seeing it close to not only confirmed my prejudice but revealed the bolder side of its character as I took in its bomb bay, gun turrets, and its functional coat of matt paint: although by this stage thoroughly dated, the Wellington had been more than a pretty face in its day.

My flying instructor, Flight Lieutenant Waddington, was a Yorkshireman; one of the quieter variety, and very much a gentleman. He had recently completed a tour of operations but, though stretched perhaps and left a little tense by the experience, he was in no sense a broken reed. I much enjoyed flying with him and tried to emulate his professional style. Only once did his calm show any sign of being disturbed – when I reported a hole in the exhaust collector ring in the engine on my side. He had that engine throttled back and the aircraft heading for base in one swift movement. No question of panic. Just a very quick reaction and a bit of breathlessness in his voice as he called the control tower for a priority landing. It was a bright sunny day, and as he explained later, hot engine gases – largely invisible in sunlight – could have streamed back from that hole, and set light to the canvas-covered wing.

Once I had mastered the basics of handling the aircraft, we linked up as a crew and began to integrate our skills under my captaincy. To begin with, we set off on a series of cross-country flights, some of which included legs over the sea, where gunnery could be practised and the wind velocity computed by measuring our drift against markers dispensed through a chute in the floor of the aircraft.

One of these cross-country flights took us via Skomer Island, off the south-western tip of Wales, to the Scilly Isles and then back to our base. With a little help from me, the last leg of the flight passed directly over my home town of Wellington in Somerset. In my next letter home, I thought I would show off a little by reporting that the garden at Wharf Cottage – my grandmother's home at nearby Nynehead – had recently been dug. But I might as well have said that the Blackdown Hills had turned bright blue for all the response this piece of intelligence produced. Clearly the family was becoming a bit blasé when it came to being impressed.

As our training proceeded, so our cross-country flights acquired a more operational flavour. By way of practice in saturating air defences, some routes included a simulated attack on an infra-red target, with dozens of aircraft attacking it within a narrow time band. In parallel, other parts of the training jigsaw – such as night flying, practice bombing and fighter affiliation sorties – were slotted into place.

In the latter, the fighter pilot tried to 'shoot us down' with his camera gun, while I did my utmost to shake him off by pulling into and sustaining the very tightest of turns; something with temporarily pinned my crew to their seats and imposed heavy 'g' loadings on the aircraft structure. Such manoeuvres were quite stimulating from where I sat, and the gunners too seemed to enjoy them as they 'shot' back at the fighter, but they were heartily loathed by the remainder of the crew. Perhaps my own enthusiasm for them might have been somewhat less had I known a little more about metal fatigue.

Life in autumn 1943 was not all work. With a weekend pass, most of us headed for London. My usual companion was Walter Stringer-Jones, the man I had pre-empted in my choice of navigator. Not only was Walter older and more experienced in life than most of us, he also knew his way around London's social life. We based ourselves at the English Speaking Union Club in Mayfair, handy for the Brevet Club in Curzon Street where we began our evenings on the town.

On one such evening we gravitated to the Embassy night club in Grosvenor Street – one of the more select establishments of its type – where Stringer wasted no time in pairing off with the sole unattached female, leaving me twiddling my thumbs. To relieve the monotony I became involved in a series of bets with two unattached American officers, whereby the odd man out – on a discreet show of coins – had to be on the dance floor with partner within sixty seconds or pay the others ten shillings a piece. If he succeeded, of course he collected a pound, as well as having the pleasure of the dance. I hasten to add that in all other respects, our behaviour was entirely decorous.

I did rather well at this game. Although frequently the odd man out, I discovered a plentiful supply of young blonde girls sitting at a table with one or two bored looking Arabs; the latter no longer in their prime, it seemed. For their part, the girls seemed quite happy to break off for a few circuits of the floor. However, it was too good to last. Before long I was called to the reception desk to take a message. In the politest terms, I was told that King Feisal was rather put out by my activities and that, although the manager wished me to enjoy myself, he would be obliged to ask me to leave and possibly to review Stringer's membership, if I did not cease my intrusions. Collapse of not-so-stout party.

I returned to my American friends much deflated, and better informed; on accepted etiquette, certainly. I am happy to say that Stringer only learnt of the matter when we met the next morning at breakfast. As it was he made light of it and was far more interested in telling me that his partner of the night before was the mistress of the Duke of—. I did my best to look impressed.

The Brevet Club was nowhere near so starchy. The beer, for instance, was kept on the customer's side of the bar and, having paid for your pint, you were handed a tankard and invited to help yourself. There was also a small dance floor, and usually a gaggle of unattached women to liven things up; as the evening progressed the women remained, but not quite as unattached. As for the atmosphere, it might have been informal but it was never less than well mannered. Its other attraction was its huge popularity with aircrew in general. You were just as likely to bump into a handful of wartime aces as you were to meet a crowd of newly commissioned pilot officers. And being a purely a social establishment – where the only membership qualification was the possession of a commission and a flying badge – everyone treated the other person on equal terms, which, as a very junior officer, I found wonderfully relaxing.

Another haunt of ours on those London weekends was Shepherds, the pub in Shepherds market; bulging at the seams whenever we looked in – usually towards closing time – and populated it seemed entirely with friendly, outgoing people. Given that the majority of the non-regulars were usually RAF, that observation might have an element of prejudice, but there was no doubting the hail-fellow spirit.

Leaflet Raid

The high point of our OTU course was a leaflet raid over enemy-occupied France, for which my own crew was targeted on the small town of Montereau, sixty miles south of Paris. Such raids took place at night, with the aircraft being despatched singly to different targets. With no instructional staff on board, and the requirement to maintain strict radio silence, each crew was very much on its own.

Outwardly, the members of my crew set about their individual pre-flight preparations on the day in question in the same matter-of-fact way as they had done throughout our training. But this wasn't just another training sortie, and they were no doubt as alive to that fact as I was. Speaking for myself, from the moment my batwoman roused me with my morning cup of tea, I was conscious that before I got between those sheets again I was in for a new experience. Although our 'target' as such was unlikely to be defended, it nevertheless lay some distance inside occupied territory, and who could say what might befall us during those hours in hostile air space? It would be misleading to give the impression that I felt in any sense windy – it was more a case of being suitably keyed up for a special event.

Apart from giving us the en route weather, our pre-flight briefing reminded us of the items each of us carried to assist our escape and evasion if we were forced down over France, and stressed the importance of not divulging anything beyond our number, rank and name, should we fall into enemy hands. Our escape kits included not only French currency, but also photographs for use by the 'Underground' in providing us with false identity papers. Posing a few days earlier for these photographs, I recall being asked to look suitably foreign and unamused. I'm not sure what we were meant to do with our faces in meeting the first part of that request, but most of us managed to look unamused at least.

We had recently practised our dinghy drill on the river at Newbury, and the chief scout himself could hardly have devised a better scale of equipment than that contained in our dinghy survival kit. Earlier that day we had sharpened our tactics against possible fighter interceptions, during which a friendly fighter had carried out a series of mock attacks from our rear quarter, whilst I did my best to shake him off. In earlier practices, the same fighter pilot had urged on me the need to tighten my evasive turns, but on this occasion he admitted that I had given him a hard time as he struggled to keep us in his sights. As we lined up to take off on the raid itself we could hardly have been better prepared, whatever the night might bring.

We staggered into the air at dusk on the evening of 5th November, hoping to a man, I'm sure, that there would be no fireworks that night on our account. I use the term 'staggered' quite deliberately, since the aircraft was most reluctant to unstick and I was obliged to use practically every yard of runway in persuading it to do so. And airborne, E for Eddie showed little more guts for the task of gaining height; making such a meal of it that I finally settled for a cruising level of just over nine thousand feet; less than I would have liked.

As we set course over Hampstead Norris, I carefully checked the heading with my navigator, turned down the cockpit lighting to an acceptable minimum, and sat back to await our crossing-out point over the coast. Meanwhile, my vision was adapting to the gathering darkness. We were on our way at last on our first operational sortie, and now that we were actually 'on stage', so to speak, what pre-flight tension I might have felt had simply ebbed away.

When the aircraft had been allocated a few hours earlier, I had not managed to talk our Flight Commander into letting me take an aircraft equipped with GEE – the latest navigation aid – which my navigator had put me up to. But I had tried, and, having drawn a blank, we now had to do the best we could using dead-reckoning navigation, astro shots and map reading; plus whatever benefit we could coax from our limited radio aids. As the flight commander had put it, 'GEE wasn't even heard of when I did my operations; you'll manage.'

At the south coast, we identified our position visually and made a small correction to our heading. Clear of land, the gunners asked permission to test their guns, and loosed off a few rounds.

About mid-Channel, we saw an aircraft coming towards us, slightly below and to one side, in-bound for England. As it came abeam of us, it was silhouetted in the moonlight against a layer of white cloud, and the intercom burst into life, as two or three voices exclaimed 'Junkers 88!' Art, my young Canadian wireless operator/air gunner (WOP/AG), adding a request to fire a burst. I cut him short with an emphatic 'No'; adding that our mission was to drop our leaflets on Montereau, not to mix it with a Junkers 88. The intercom lapsed into silence. Quite apart from the need to maintain the aim of the mission, I could hardly imagine that a fleeting burst of .303 would be likely to cripple the JU 88, and I certainly would not have fancied matching a particularly docile and dated Wellington Mk 1c against an aircraft which numbered among its roles that of night fighter.

We looked without success for a landfall on the French coast, in spite of my airbomber lying prone in his bomb-aiming position. And although we later made one or two alterations of course en route to the target based on dead-reckoning calculations, I gradually formed the view that we would be extremely lucky to pinpoint Montereau, given the deteriorating visibility and the lack of distinctive ground features in the target area.

Indeed, after three hours, and having reached our estimated time of arrival at the target, the best that Leo, my navigator, could say was that we were somewhere in the vicinity of Montereau. Given the passive nature of our load, I decided that that would have to do. We released our leaflets, and with the aid of a flare, photographed what lay below.

Our leaflets disposed of – hopefully not all of which would be eaten by the local cattle – my airbomber went aft to top up the engine oil tanks. This involved handpumping oil from a fuselage tank to both of the engine nacelles in turn; the whole operation taking about ten minutes. This was standard procedure for the Pegasus XVIII engine after the first three hours, after which it was repeated every hour. An overriding requirement was that if the oil pressure warning light came on, the handpump had to be operated until the light went out.

The return flight proceeded uneventfully for the first hour, except for the cloud cover beneath us progressively increasing as we made our way north, which did nothing to help our navigation. At this point the oil tanks were once again due to be topped up. But when I indicated this to my airbomber, he demurred, complaining that he felt sick. Not wishing to show him up unnecessarily, I switched off my microphone, removed my face mask, and leaning across to where he lay slumped in the second pilot's seat, I ordered him to get aft and start pumping. At that he hauled himself out of his seat and disappeared towards the pump. Short of taking someone else off an active task, I had no option.

After some minutes with no word from the airbomber, I sent the wireless operator to see how the pumping was going. In less than a minute, the wireless operator reported that he personally would be doing the pumping and would return to his set when the job was done.

No one who has been airsick, or close to it, would underrate its power to test a man's resolve, but it is potentially threatening when it disrupts the efficient working of an aircrew on operations. As the flight continued, my reaction to my airbomber's lack of determination was one of mounting disgust. He was virtually a passenger from that point onwards; spending a good deal of his time, where the wireless operator had found him, draped over the Elsan. Later in the flight, the oil pressure warning light came on on two or three occasions, and the wireless operator no doubt felt a bit like the proverbial one-armed paperhanger as he tried to cope both with pumping of the oil and the operation of his set.

Using the radio beam equipment in the latter stages of our return, we eventually homed overhead to Hampstead Norris (or was it Harwell?) and let down on instruments through a thick layer of cloud. As we began to emerge from the ragged cloud base, I realised I was becoming disorientated, and for a second or two I had to use every ounce of will to keep control my senses and to fight off a feeling of near panic. With only a partial view of the airfield lights, I had begun to align the wings with the reflection of the flarepath in the sloping glass of my side window instead of on the flarepath itself. The urge to follow that false 'horizon' was so compelling I knew I must get rid of it, and quickly. My reaction was deliberately to re-enter the cloud, settle myself once more on the artificial horizon, and this time to let down with my seat lowered and with my eyes riveted on the instrument panel until we were completely clear of cloud. It worked, and I had no further problems. But the six and a half hour flight, most of it in enemy airspace, and the aggravating discovery of a weak link in the crew had obviously left me a bit spent.

On landing we were debriefed by the intelligence staff. Not that we had much to report, other than to say we had been obliged to drop our leaflets on dead-reckoning navigation, and that outbound we had seen a JU 88. At that stage we could only hope that our vertical photograph would tell us just where we had dropped our load.

By the time we reached the Mess and sat down to our supper of bacon and eggs, my airbomber seemed to have largely recovered as he entered brightly into the conversation; something which added to my suspicion that his sickness might be fundamentally a nervous problem. It did nothing to lighten my mood and later, as I tried to sleep, I found myself going over the events of the sortie and reflecting on what action I should take.

The next day I took the painful step of reporting adversely on our weak link. It was an unpleasant task, but one which I could not duck, knowing that on operations proper – as opposed to a leaflet raid over rural France – there would be even less room for a passenger. Any shred of doubt I might have had about taking that course was quickly removed when, first my navigator and then the remainder of the crew, spontaneously expressed the wish not to fly on operations with a man who could be so incapacitated by air sickness.

In the closing weeks of our course, our minds turned increasingly to our next posting and which of the heavy bombers would be our next machines. We were distinctly unimpressed with accounts of the Stirling, which apparently could not struggle much above 14,000 feet on its bombing run, and was consequently a favoured target of the German Anti-Aircraft batteries. Stories about the Halifax, although not nearly as damning, were not all that encouraging either: apparently it was also a bit short on altitude during the bombing run – being limited to 18,000 feet or so, as compared with the Lancaster which bombed from heights between 22,000 and 27,000 feet. Not surprisingly, our dearest hope was to be posted to a Lancaster unit, and if that was not possible, that we would at least find ourselves on the Halifax – with improvements in its performance hopefully in the pipeline.

But a week or so before the end of the course, our options were suddenly widened. Out of the blue came a request for volunteers to serve in India, where they would fly American-built Liberators on heavy bomber operations against the Japanese. A new squadron was to be formed and conversion training would also take place in India. Whole or part crews, less airbombers, were required and the volunteer pilots would become aircraft captains on successfully completing the conversion course; as distinct from a UK Bomber Command squadron where, it was rumoured, one could expect to serve for some time as a second pilot.

I was greatly taken with this new opportunity. More travel; an immediate captaincy; an aircraft with a handsome reputation. What more could one ask? Regrettably, Leo my New Zealand navigator, already well into his thirties, was opposed to another long sea voyage before starting a tour of operations, so it would mean losing him, but Tex, my rear gunner, and Art, my WOP/AG, shared my enthusiasm. So in spite of having to say goodbye to Leo, the rest of us volunteered and were accepted. I was pleased to find that Walter Stringer-Jones had also taken the plunge.

So, despite the downbeat note introduced by my airbomber's behaviour on our operational sortie, I left 15 OTU with my morale riding high; boosted by the thought of serving in India – which had so impressed my father when he was my age – and the prospect of a captaincy on the pick of the Allies' long-range bombers.

To India by P & O

Between my operational training and embarking for India, I spent five weeks at home, as one extension of leave followed another. My mother was more than happy with this arrangement, as long as the RAF kept me supplied with ration cards, and this extra leave also gave my father and me the opportunity to grow closer in our adult relationship. It was mid-December by the time I reported to an embarkation holding unit in Blackpool. But from that point things moved fairly quickly. In less than a week our convoy had put to sea.

For us, Christmas Eve 1943 was a clear, starlit night: with the Rock of Gibralter plainly visible as we slipped quietly through the straits into the Mediterranean. Our convoy could equally well be seen from Spanish territory of course, and one could not help wondering what sort of reception might lie in store as we made our

way to Port Said. The Axis forces had been driven from North
Africa seven months earlier, and Sicily had also been captured by
Allied Forces. But the island of Crete, with its airfield, remained
in German hands and one had to assume the presence of enemy
submarines.

Passing through the Mediterranean we were subject to General
Eisenhower's orders in his capacity as theatre Commander-in-Chief.
This meant no drinking of alcohol, and in those troopships with a
shortage of bunks (in effect all troopships, I imagine), non-comba-
tants having to make do with mattresses on the cabin floor. As a
combatant, I accepted this latter arrangement without reservation,
but from the reaction of one or two non-combatant officers in our
cabin – Flight Lieutenants approaching early middle-age – you
would have thought their world had been stood on end.

Before hostilities, our ship was in service with the P & O Line,
and if I had to put a name on her, I would say she was the SS
Strathaird. In contrast with the *Queen Elizabeth*, in which I had
returned from Canada and which had yet to enter fare-paying passenger service, our P & O ship had a mature, lived-in atmosphere.
In the dining saloon at least, she seemed to have retained many of
her peacetime stewards – mostly Asian – who bustled to and from
the galley in a great display of willingness and expertise; their manner
alive, attentive; their hands and forearms festooned with plates. The
food too was more than passable, although some of the courses
became a bit monotonous: supplies of Brown Windsor soup and
Cabinet pudding being inexhaustible apparently.

Our convoy called briefly at Algiers, where we put one or two
people ashore by launch. But the nearest I got to sampling the
delights of the Casbah was listening to some old sweats from
the Indian Army recalling their escapades of earlier voyages, as they
gazed wistfully from the ship's rail.

We also had a party of Wrens on board, and a few opening moves
were made by way of courtship whilst we were still west of Suez.
The pairings which lasted, however, were to come later; in the
Red Sea and beyond, where passions steadily rose with the higher

temperatures, the change to more revealing clothing and the virtual absence of serious distractions.

Sixteen junior officers shared the cabin I was in. We were a mixed bunch – twelve aircrew, of whom nine were RAF and the remainder RAAF; two RAF ground duties officers; and two Royal Navy deck officers. One of the RAF aircrew had seen service as a Hurricane fighter pilot escorting convoys to Murmansk. When an enemy aircraft was sighted, these brave men – all volunteers – were catapulted from the decks of merchant ships and, given that they won the encounter, they afterwards ditched or baled out close to an escort vessel, hoping to be rescued before perishing in the Arctic waters. This particular man was a Scot, and had stitched a piece of his clan's tartan to the front of his flying helmet, and whereas in most squadrons I dare say this would have been regarded as a bit flashy, one felt that a man of that calibre could be forgiven a touch of flamboyance.

Perhaps the man with whom I would have been least prepared to swap jobs was Flight Lieutenant Wilcox, a navigator. He was on his way to serve with General Wingate's Chindit force as an RAF liaison officer, where his job would be to accompany a guerrilla warfare column deep behind Japanese lines, guiding in the Dakotas and other resupply aircraft, so that their parachuted loads would be delivered accurately and on time. Wilcox was a gritty Yorkshireman of rather spare build, frequently wound up about something, and generally edgy. I wondered how he would cope when the pressures were really on. In the event he did very well. And although he was finally evacuated from behind enemy lines for reasons of poor health, it was not before he had survived some bloody hand-to-hand engagements with the Japanese, as well as doing a professional job in his specialist role.

We junior officers were not completely without some responsibility for the ship's routine. Each of us was allocated a small batch of airmen, and was answerable for their personal welfare throughout the voyage. Among my batch was another Wellingtonian; Clem Holley, who had been a very useful three-quarter in the town's rugby team. It seemed strange to begin with, having responsibility for a man I had so recently looked up to for his sporting prowess and who was much my senior in years. Not that either of us showed any outward embarrassment at this turn round.

Most of us, I imagine, had expected the enemy to give us some stick whilst we were in the Mediterranean, but it was not until we were south of Crete, that he showed his hand. At that juncture we hurried to our lifeboat stations in response to an air raid alert, but even then – while we could hear aircraft operating in the vicinity, and our surface escorts were dashing about as though things were about to happen – the convoy itself was not attacked. And I'm glad to say it remained that way.

Our arrival at Port Said was greeted by shoals of small trading boats, their occupants eager to relieve us of our pocket money in exchange for leather goods, trinkets, hats and basketware etcetera. Lines were hurled up from below with shouts of 'Johnnie, catch!' And with the means of passing goods and money established, the bargaining began. It was all good humoured, brisk and amusing; and how refreshing it was to have contact once again with people on shore.

This trading with the bumboats was accompanied by the wholesale dumping of 'Simpsons Bowlers' into the harbour below. Once this got going, it caught on among the troops like a bush fire. These huge moulded sun topees, a standard issue to all ranks, were ungainly beyond belief, and after a few days in India, few of us had not acquired something lighter and more compact from the local bazaar. What the troops had done was clearly an act of defiance, but as the ship's brig could hardly have held five hundred men, what else could OC Troops do but turn a blind eye? Meantime, it was almost impossible to keep a straight face at the sight of all those pristine regulation topees bobbing about in the murky waters of the harbour.

At about this time, we went into tropical rig, and any doubt as to who had been in the tropics before was instantly dispelled. The old sweats had trim, individually tailored uniforms and, in many instances, the residue of a tan to build on. The rest of us, wearing regulation khaki drill, were characterised by baggy shirts, over-long shorts and lily white knees. Admittedly there were some exceptions to these generalisations – among the veterans, a handful of middle-aged Indian Army colonels, whose shorts almost touched the tops of their stockings just below their knees, and who looked like museum pieces; and among the newcomers to the tropics, the Wrens who, if anything, looked ever dishier than before, and whose tailors had somehow found the nerve to flout the prudish regulations of the day, by producing garments which flattered rather than detracted.

Soon we were passing through the Suez Canal, with everyone on deck savouring the proximity of land, and those of us new to the scene avidly taking in each fresh item of exotica. Not that the latter were especially numerous, given the general absence of habitation along the banks. But – Port Said apart – when you've looked out on nothing but ocean and ships for a week or so, a camel train, some date palms and a few hoary old Arabs dressed in colourful nightshirts can make quite an impact. The troops, knowing the Wrens were out on deck in force, added spice to the occasion by successfully goading a passing Arab to reveal his manliness – which brought a huge cheer from the troops and not a few blushes from the sun deck. To judge from that Arab's defiant and full-blooded response it was not the first time he'd engaged in a 'dialogue' with a passing troopship.

As we made our way east of Suez, through the Red Sea and across the Bay of Bengal, the entire mood of the ship seemed to brighten. We had left one war zone and had not properly entered the other. The sun shone all day; the nights were pleasantly warm; we had long since found our sea legs; and the fruit and vegetables taken on at Suez had relieved the monotony of the ship's menu. Boxing tournaments were organised; the chief Wren was probably mindless with concern for the chastity of her girls; and we had a ship's concert – at which, after we had sung *Good Night Ladies*, OC Troops, obviously tiddly, strutted up and down, swagger stick at the slope, telling us of his young days and of his reputation as a stage door johnnie.

The sea too came to life. Schools of porpoises shepherded us for long periods at a time, undulating gracefully in and out of the water, as they kept station a hundred yards or so off our bow. Closer in, small shoals of flying fish occasionally shot forward from the bow wave to take flight for ten or twenty yards, their silvery bodies glistening in the sun, before they lost lift and flopped back into the water. Neither did the fascination of the ship's rail disappear at night, when our wake glowed with an uncanny phospherescent light and the soothing noise of cascading water prepared us for bed.

Weeks of inactivity inevitably led to some horseplay, to which our cabin was no exception: people being singled out to provide the butt for corporate humour. First, Stringer-Jones's colourful plaid pyjamas, together with his Oxford accent, were taken exception to, and just as he was about to turn in, he was seized and dumped in a bath of cold salt water. He put up some resistance, but realising the odds were against him, soon gave in and, taking it in good part, gained noticeably in popularity. Next to receive attention were the naval officers, who it was felt should be debagged, partly because they were always so irritatingly cocksure, and also because it satisfied air force prejudice. But in this, we had reckoned without the navy's tactical skill. As soon as our web-footed friends saw what we had in mind, they promptly upstaged us by removing their trousers and handing them over. What spoilsports.

The final twist of this schoolboyish phase was a bid to apply boot polish to the backside of our rather priggish education officer. But the group was split over this idea, which some felt exceeded the bounds of good humour. When he resisted – which he did with quite unexpected ferocity – our ranks wavered, and he was let off. The thing which really scotched that affair was the positive dissent of not only died-in-the-wool supporters of the underdog like Wilcox but, surprisingly perhaps, Stringer-Jones. On that discordant note, corporate horseplay was abandoned.

Within a few days the youngest member of our cabin totally restored our unity; albeit accidently. This wide-eyed young officer, who had been drooling over a pretty young Wren since Suez, suddenly burst into the cabin in a state of wild excitement. Unable to contain himself, he cavorted about like a Morris dancer, exclaiming loudly that they had just made love in her bunk. For a variety of reasons everyone of us resented this unseemly announcement: we younger ones perhaps, because the blighter had succeeded where we had not; the not so young, because he should carry on so about it. And Stringer-Jones and like-minded, because, by his open proclamation, he had broken confidence with the young lady. Poor chap, he seemed quite taken aback that we were not queuing up to slap his back. At a stroke, the rest of us had forgotten our differences.

India

We had awaited our landing at Bombay with some eagerness, although I suspect few of us making our first visit to India were prepared for anything quite as exotic. One of the first things to catch our attention as we entered the harbour was the striking Gateway-of-India monument, erected to commemorate the visit of King George V and Queen Mary in 1911. But for me, the real impact came once we were on shore and from things essentially Indian.

So many things were present in the extreme: the hordes of people; the flamboyant colours; the poverty; the wealth; the pungent aroma of strange spices; the sensuous smell of sandalwood; the stench from the open gutters; the persistence of the street beggars and the repugnant disfigurement of those they paraded to evoke our pity; the sacred cows ambling aristocratically and unattended on the city streets and pavements; the blare of a dozen taxi horns as their owners jockeyed for position; the wailing lilt of the streetsellers proclaiming their wares; the statuesque deportment of a housewife walking by with a large brass water container delicately balanced on her head; the majesty of a huge elephant, adorned with colourful trappings and moving with such quiet grace; the fierce pride of the Pathan watchman at our Worli transit camp. All this and much, much more.

We spent several days in Bombay, blowing off steam and generally enjoying ourselves. By day, we swam at the Willingdon Club and attended the races, while at night we indulged ourselves in the restaurants and bars of the better hotels, as well as satisfying our curiosity by winkling out more exotic places of entertainment.

At the racetrack, Flying Officer 'Mac' McCormick – whose father had once cleaned out the local bookie in his native Australia – was only too willing to guide us in the mysteries of the tote: which ensured that we enjoyed the day without actually losing our shirts. But personally I found the dazzling assortment of spectators equally as interesting as what the horses were up to, the Parsees especially catching my eye. Members of a religious sect whose ancestors left Persia many centuries ago, they have resolutely preserved their identity. The young Parsee women, with their pale olive skins, aristocratic good looks, and their slender bodies wrapped in delicately coloured saris, were the epitomy of elegance and serenity.

At night, wherever we went, we seemed continually to be bumping into familiar faces; not the least of whom was OC Troops himself. It appeared that there was no one from the ship who wasn't bent on letting go. It is nice to think this brief pause in Bombay was deliberately built into our itinerary by a thoughtful authority to allow us to get rid of the tensions of a month in close confinement. But whatever the reason, it made a welcome and invigorating change.

Assembling at the Victoria Terminus railway station for the next stage of our journey exposed yet another layer of Indian life. It began with the building itself. With its handsome architectural features, its striking symmetry, its huge crown-like dome, and the inspiring pose of the solitary figure which capped the whole, it looked almost too grand for such a mundane role. It was as though the architect had deliberately set out to put St Pancras itself in the shade. If so, the result was a close-run thing.

Inside the station, the magnificence of the entrance corridor and main hall was lost amid the clamour and colour of a sea of humanity, as people jostled to make their way to and from the waiting trains. Adding to the feverish activity were the traders, noisily peddling their refreshments, souvenirs and betel nut. Neither was there any shortage of little boys eager to relieve us of a few annas by shining our shoes or who, along with their even smaller sisters, just thrust out a tiny palm and pleaded for backsheesh. By the time our train pulled out and we had left behind the general hullabaloo, it was a relief just to sit back and drink in the changing landscape.

A word or two about the betel nut sold by the traders: this is a mild drug (contained in the nut and the leaf of the betel palm) used by many Indians, much as cigarettes are smoked in other countries, the main distinction being that the betel nut is chewed and the habit is confined mainly to the lower orders. Happily, slaves to the betel nut do nothing to endanger the health of others, but they do leave their trademark. As they chew the nut their mouths become stained a rich blood red, and since they frequently clear their throats, the result is that railway platforms and street pavements are often copiously streaked with bloody stains: an alarming sight to begin with.

To proceed with our journey. At that time of year, approaching the end of January, we were in the middle of the dry season, with the climate at its best. The days were pleasantly hot, with temperatures often in the low eighties, while overnight temperatures demanded a blanket or two: a combination which made for a good night's sleep, and left us the next day – in spite of the cramped conditions of a railway compartment – alert, and keen to absorb our new surroundings. Spectacular sunsets became a daily feast, and each day brought some special novelty among a host of new images as we threaded our way across India: stopping at a station one night with the air still warm, we were treated to the extraordinary sight of a bush lit up by hundreds of fireflies, glowing like miniature lights on a Christmas tree; on another occasion it was the mischievous, uninhibited behaviour of a troop of monkeys which held our attention.

From Bombay, we went almost a thousand miles east to Salbani, an airfield on the Bengal Plain, sixty miles west of Calcutta. And it was there that we began our Liberator conversion course. The date of my first flight was 3rd March 1944; ten and a half weeks after leaving the UK. There was a great deal to admire about this aircraft, and for reasons I explain later, my crew and I took to it immediately.

On advice from the old hands, almost the first thing I did on arrival at Salbani was to order a pair of calf-length boots as a precaution against snake bites. Some of the really venomous Indian snakes, like the silver Krait, are quite small; and the danger lies in treading on them in the dark, with the snake sinking its fangs into one's lower leg purely as a reflex action. My boots were ready within forty-eight hours, made-to-measure in the local village. After that, I could at least walk between the Mess and my bungalow at night, and occasionally to the camp cinema, without carrying my heart in my mouth the whole time. But to be honest, my fear of snakes never really left me throughout my stay in India.

At Salbani, even within the comparatively domesticated nature of an RAF station, life could still produce the occasional exotic touch. There was, for instance, the orderly officer's announcement at an evening cinema show that a tiger had been seen on the camp, and his accompanying advice that we should return to our quarters in groups and not singly. No one, but no one, questioned that advice. For good measure, since one of the Mess servants kept a goat tethered near my bungalow, I slept that night with my .38 Smith and Wesson on the bedside table, loaded and ready.

Inside a few days, I realised that wild animals were not the only strange species to be found on the camp. The rear of my bungalow backed on to that of a senior member of the staff, and as he went through his ablutions a matter of yards away, he held animated conversations with himself as a matter of routine. I made no conscious effort to catch what he said, but like it or not, I was soon aware that he quite approved of himself. The other members of staff seemed well acquainted with his habit, and agreed that he was a bit 'round the bend'. But they did not seem unduly bothered by his behaviour. As far as they were concerned, the risk of developing such characteristics was an occupational hazard of an overseas tour, and they suggested I would meet others like him.

There were also some thoroughly normal and very agreeable members on the staff of the Liberator conversion unit. Settling into my room on the evening of my arrival, the relaxing strains of Glen Miller's 'Moonlight Serenade' drifted in from a wind-up gramophone in the room next door, and I was quickly in amicable conversation with the owner of that splendid instrument – one Flying Officer Johnnie Hollingsworth, the motor transport officer.

Thanks to Johnnie, not only did I continue to enjoy the soothing effects of his gramophone each evening, but I had soon mastered the rudiments of driving the Ford and Chevrolet three-ton lorries which made up the bulk of his fleet. Most evenings after dinner we set out for the airfield, and on the quiet, deserted roads linking the aircraft dispersals, he took me through such techniques as double declutching and the use of the accelerator when changing down. None of this was relevant to my job at that time, but the opportunity of learning to drive seemed altogether too good to miss. (It may seem a little odd by today's standards, but along with many other qualified pilots, I had still to drive a car at that stage.) And although I would not claim that mastering a three-tonner was quite as exciting as learning to fly, it at least made a change from an evening at the camp cinema.

We were not done with travel. At the end of March, the conversion unit moved from Salbani to make way for 356 Squadron; a Liberator heavy-bomber unit which had formed at Salbani two months earlier, and which needed extra facilities as it was fleshed out. We had barely reached the mid-stage of our conversion course by then, and as unit property, we were parcelled up and despatched along with the rest of its belongings to RAF station Kolar, close to Bangalore in the South Indian state of Mysore.

Apart from the aircraft, some essential items of support equipment and several key members of staff, the entire move was made by rail. The journey alone took us three to four days; which makes the break of only fifteen days in our flight conversion course quite an achievement. As with the rail journey from Bombay to Salbani, we again covered about a thousand miles; and viewing India once more at a leisurely pace and from ground level, merely added to my fascination for its variety and mystique. It was during this journey that I added the delectable pawpaw to my growing list of new fruits: the mango having already won me over since our arrival in Bombay.

Although we recommenced our Liberator training at Kolar with commendable speed, other aspects of life at our new unit got away to rather a poor start. Domestic administration was lacking, especially, it seemed, in the matter of sorting out complaints from the airmen before things got out of perspective. Fairly soon, Kolar had something approaching a mutiny on its hands.

The Unit Commander reacted by beating everyone about the ears with the rule book (King's Regulations). All ranks, staff and students alike, were taken fifteen to twenty miles from the camp in three-ton trucks, and made to march back in the heat and dust of the day: the

Unit Commander himself occasionally taking his place at the head of the column. Perhaps things had reached a point where the whip needed to be cracked, but it was a sad experience. And although communications within the unit improved after that, the whole episode left a bad taste, making the escapism of flying even more welcome than usual.

Within days of resuming my conversion course I went down with bronchitis, which became so persistent that I was finally admitted to the British Military Hospital in Bangalore. With rest and regular medication I made progress, but recovery was a slow business, and not long before my release, one of the medical officers – after hinting that my recovery was unaccountably slow – put it to me that if I found the prospect of operational flying too much for me, there would be no shame in my admitting it. I practically leapt down his throat as I assured him I had no such anxiety. In posing that question, he may or may not have deliberately used a psychiatrist's ploy, but I was so incensed by this slur, and the thought that those in command at Kolar might have been advised of his 'suspicions', that I've no doubt my recovery was stimulated by that episode.

A week later I began two weeks' sick leave at Ootacamund, a beautiful hill station near the summit of the Nilgiris, known, even among the British, as 'Snooty Ooty' because of its pukka ways. The officers' leave centre where I was accommodated, was based on Arranmere Palace and was run in some style by a titled British lady: Lady Shaw, if my memory is correct. But, although we made use of the palace facilities for dances and the like, most officers, including myself, were billeted in one of a number of dormitory houses separate from the palace itself; each run by a British hostess.

Unfortunately, our particular hostess had a vile temper, which was invariably on a short fuse and which got even worse with drink. Most of us coped with the problem by keeping out of her way, but it was inevitable that someone would stand up to her from time to time. During my stay she met her match in Wing Commander Freddie Snell, who saw her as a tyrant, and chose to tell her so one evening in front of the rest of us. For most of us, I dare say, there was more strain in witnessing their verbal scrap than there would have been in a hard week's work; glad as we were to see her get her come-uppance. Our hostess's immediate reaction, apart from turning bright scarlet, was to tell Freddie to pack his bags and leave the next morning. Overnight, however, Freddie's expulsion was reconsidered, and the two settled down to a watchful truce.

Forgetting the brittle atmosphere of our dormitory house, there were many convivial and pleasant moments at Ootacamund. At the palace dances, there was the delightful and amusing company of Alison Hope and Alison Lothian, the daughters of senior members of the British administration. And the lush scenery and the cool climate of the hills, in themselves, were a balm after the dust and heat of Kolar. On balance, I decided that whoever coined the phrase 'Snooty Ooty' probably did so out of envy as much as anything. When I took the train to Kolar, I felt thoroughly rejuvenated and eager to get on with my conversion course.

Things had not stood still in the two months since I had last flown. My contemporaries had completed their course and had already returned to Salbani to join 356 Squadron. Worse still, my own crew had been adopted by a member of the instructional staff who was returning to active operations, and was about to depart for another of the heavy bomber squadrons in Bengal. As a result, I was given an entirely new crew, with whom I completed my training in early July.

It was a blow losing my original Liberator crew, especially Art and Tex, who had been with me from my days on the Wellington. But if I had to lose them, I could hardly have done better by way of a new crew. Individually they were bright and spirited. And, as regards their origins, it would have been difficult to assemble a more diverse and interesting assortment, drawn as they were from all parts of the Commonwealth and beyond.

Jack Clarke, my second pilot, came from New Zealand and belonged to the RNZAF. A cheerful, bouncy character by nature, Jack had a sardonic wit – mostly aimed at authority – which frequently gave us a laugh, and in moments of stress helped to relieve the tension. Clearly, the aim of any pilot – having gained his Wings – is to captain his own aircraft, and for such a man to serve in a supporting role for any length of time and still maintain his professional edge and enthusiasm would be a difficult challenge. To his credit, Jack gave me unstinting support in the months ahead.

Jock Hunter, our navigator, was a Scot on temporary release from the Metropolitan police force in London. A big man in every way, Jock's calm, unflappable nature was to prove a great asset. When things went wrong, which occasionally they did, he had the capacity quickly to pick up the pieces and get on with it: an important quality in most circumstances, and on operations sometimes a vital ingredient.

Terry Malcolm, my wireless operator, was a member of the Royal Australian Air Force (RAAF). Tall and thin with a lantern jaw, a freckled face and carrot-coloured hair, Terry was no yes-man. Certainly you had to earn his respect, but that bridge crossed, he was intensely loyal and a pillar of strength within the crew. And in the air, it would have been difficult to find a more cool and professional person.

Perhaps the most colourful character in my new crew was Chris Matthews, our second wireless operator-cum-beam gunner. Chris or 'Chile' as he was popularly known, was a British Latin American Volunteer from Valparaiso, who proudly wore the letters BLAV on the shoulder of his tunic. An ebullient young man with Latin blood in his veins, an aquiline nose and a luxuriant head of black wavy hair, Chile was cheerfulness personified and frequently regaled us with his good-natured wit.

'Smithy', the mid-upper gunner, was a Canadian and a member of the RCAF. A big, vigorous young man with a red, shining face and a mop of blond hair, he was very open and forthright; occasionally bordering on the tactless perhaps, but energetic and full of enthusiasm for his job.

Arthur ('Red') Rigby, on the other hand, was the soul of discretion. He was my flight engineer who, over enemy territory, would combine those duties with manning one of the two free-mounted guns at the beam position, aft of the wing. He would also share with Chile the job of taking photographs with the hand-held camera stowed at the beam position. Red, who was the second Scot in the crew, derived his nickname from the colour of his hair. But there was nothing in the least fiery about his temperament: on the contrary, he was a quiet, conscientious man who appeared thoroughly content poring over his fuel charts and advising me from time to time how much petrol we had to play with.

The remaining member of my new crew, at that stage, was Ted Holmes my rear gunner. A family man in his late thirties, and very much the veteran of the crew, Ted was a short, rather wizened man with a soft Norfolk accent and the ruddy complexion of a farmer. He was also one of nature's gentlemen, who with his conscientious devotion to his job, and his quiet, mature ways quickly gained everyone's respect.

From Kolar we were first posted to 159 Squadron at Digri, about twenty miles north of Salbani. And although this would have meant that I would not, after all, rejoin the friends I had been with since beginning my operational training in England, I was quite pleased at the thought of serving under Wing Commander Blackburn, the 159 squadron commander, who was something of a legend in South East Asia Command (SEAC) as a determined and innovative leader.

Blackburn's special skill lay in his ability to squeeze extra long sorties out of his Liberators, whilst doing so with the minimum reduction in weapon load. His method of apparently achieving something for nothing included dispensing with some of the standard protective armour and armament in order to carry extra fuel. Not that such moves resulted from a mere chancing of his arm: on the contrary, since his squadron specialised in night operations, and his tactics relied on stealth and audacity, they were very much in the nature of a calculated risk.

The benefits of Blackburn being allowed to modify his aircraft in this manner are best illustrated by one of 159 Squadron's operations, during which – in October 1944 – twelve aircraft successfully mined Penang Harbour in a non-stop flight of seventeen hours, involving a round flight of three thousand miles. To lend some scale to this achievement, a Liberator not specially stripped down would have a maximum working endurance in the order of fourteen hours.

However, having built ourselves up to serving under this legendary figure, on reporting to Digri we found we were not wanted. It appeared that the personnel branch had meantime changed its mind, and we were redirected to Salbani to join 356 Squadron without as much as unpacking our bags.

4   *Bomber Operations*

356 Squadron – Backcloth to Operations

A posting to 356 Squadron might have lacked the status of joining the somewhat elitist ranks of 159 Squadron, but it was a very pleasant reunion. Arriving at Salbani was like rejoining one's family. I had, after all, caught up with Walter Stringer-Jones and other friends who had left me behind at Kolar.

To put 356 Squadron into perspective, it was at that stage (mid-July 1944) one of three RAF Liberator heavy-bomber squadrons based in India. The others were 159 Squadron, which had seen operational service in the Middle East before arriving in India in October 1942, and 355 Squadron which had formed at Salbani in August 1943 and with whom we now shared the Salbani airfield and station facilities. All three squadrons came under the command of Headquarters RAF 231 Group Calcutta, which exercised control through the respective Wing headquarters at each airfield.

The Air Officer Commanding (AOC) RAF 231 Group, Air Commodore Mellersh, also had under his command at that time two Wellington bomber squadrons – 215 and 99, based at Jessore. These squadrons joined the Air Order of Battle as Liberator squadrons later in the year, after completing their conversion in August and September respectively: both squadrons certainly being in action again by November 1944, and possibly earlier. So it will be seen that before the end of the year our AOC had five Liberator bomber squadrons at his disposal.

The Liberator bomber scene in India would not be complete without reference to the aircraft of the 7th USAAF Bombardment

Group, which comprised four squadrons: the 9th, 436th, 492nd and the 493rd. Both the 7th Bombardment Group and Headquarters RAF 231 Group came under the overall control of Headquarters Strategic Air Force (SAF) Calcutta, an integrated British/American Air Command formed in December 1943, and in which – by July 1944 – the RAF played the dominant role.

Turning to the general situation: the Japanese still occupied the whole of Siam, French Indo-China and the Malayan Peninsula; and although they had been pushed back from the border with India, they also remained largely in possession of Burma. The immediate objective of the British Fourteenth Army under General Slim was to retake Burma, and the job of SAF was to assist him in that task. The Japanese had long lines of communication and relied heavily on ports, inland waterways and railways for their strategic supplies: the infamous Burma-Siam railway, built under pitiless and degrading circumstances by prisoners-of-war and forced indigenous labour, being one of those lines of communication. This total transport system, together with forward supply dumps and fighter airfields, was to be the primary target of the Liberator squadrons in the months ahead.

Of special significance to the Liberator squadrons as they entered the period of the South West Monsoon in June/July 1944 was Lord Louis Mountbatten's decision – taken on his arrival in the theatre the previous December – that air operations were henceforth to be conducted the year round; thus breaking with past practice, whereby during the SW Monsoon they were largely suspended (by both sides) in deference to the extraordinary violence of the thunderstorms.

As to 356 Squadron itself, and in particular the make up of its aircrews; this was an interesting blend in a number of ways, and a blend which, even in my time, was to change fairly radically as the original crews were posted on completing their operational tours.

When the squadron was formed, its root source of crews was 15 (Wellington) OTU at RAF Harwell via the Liberator Heavy Conversion Unit in India. At that stage, apart from two RAAF captains and a sprinkling of crew members from the Commonwealth and beyond – my own crew having an unusually generous sprinkling –

the aircrews were made up mainly of RAF personnel. From November 1945 however – with the closure of the Liberator conversion unit at Kolar and its re-emergence as 358 Squadron – we came to reply on the Liberator conversion unit at Boundary Bay on the Canadian west coast for our supply of crews. As a result, by the time I left the squadron at the end of May 1945, the majority of the aircrews were RCAF.

In due course, contact with the new crews confirmed my view that Canadians make excellent airmen, with their robust outlook and their generally unflappable temperament. They were also to bring with them new social habits with which they enlivened Mess life. It was not long before the less staid of us 'oldies' joined them on our knees in a corner of the ante-room to play craps; the dice game I had first encountered during my training in Canada, when I had thought it a particularly reckless form of gambling, but which at Salbani I found myself good and ready for.

The squadron blend was also interesting for the disparity in ages among the aircrew. Although most, like myself, were in their early twenties, we had a few members who were well into their thirties. Some of the latter, like Walter Stringer-Jones and Les Campnett came from reserved occupations, and need never have put on a uniform but for their patriotism. We were all volunteers of course, but these 'veterans', as we regarded them, had taken enough of life's knocks to be more conscious than most of us of the dangers we faced. As they got further into their operational tours there were occasions, I'm sure, when they had to dig much deeper into their reserves of will or determination – call it what you like – than us untried, cocksure young men; particularly when replacement crews failed to materialise on time and, as a result, our operational tours were arbitrarily extended.

But to return to the position in mid-July 1944: 356 Squadron had recently emerged from its shake-down training phase, and was now firmly on the Air Order of Battle. However, although it had an operational meteorological sortie to its credit, the squadron had still to perform its first operational bombing sortie.

Wing Commander Hugo Beall, our Squadron Commander, had

by all accounts worked the squadron hard in training, and although the tactics for each type of attack were to some extent still being fine-tuned, the leaders of the main and subsidiary formations had already been selected, and individual reputations were beginning to take shape.

As to Hugo's leadership in the air, this had still to be tested on operations, but it appeared to be extremely sound, and he was continuing to demand from everyone the highest standards of professional conduct. There had, however, been one episode which, on the face of it, appeared to reflect adversely on the navigational skill of his crew, and which, in turn, had placed some of his sub-formation leaders in something of a dilemma. Returning from a navigation-cum-bombing training flight in squadron formation, Hugo had led the squadron progressively further off the intended track, until the error assumed such proportions that one or two of the sub-formations had chosen to break away and return independently.

By the time I arrived at Salbani, the aggravation caused by this temporary breakdown in formation discipline had been largely forgotten. Nevertheless, it did seem rather odd that with the same navigation aids available to each aircraft, it was Hugo's crew which should have been so late in appreciating the track error; and I could not help feeling there must have been more to it than was being passed on to me. The missing factor was to become all to clear to me later.

On 27th July, the squadron despatched seven aircraft to carry out a formation bombing attack on Yeu, about seventy miles north-west of Mandalay, and while it was not quite a copybook raid, as a squadron we had at last broken our duck. As to the detail of that raid: on the run-in to the target one aircraft accidentally released its entire bomb load, and the remaining aircraft abandoned their attack because the target was obscured by cloud. However, not to be denied, our aircraft then flew to Kongyi and successfully attacked a supply dump.

Being a new crew, we were prepared to make do with training sorties initially. But after a while we had had our fill of these, as we continued to polish up our navigation, bombing, formation and fighter-evasion techniques. And when by the end of the second month, I had not even flown on an operation as a second pilot, I informed Ted Needs, Hugo's adjutant, that I was about to apply for a posting to a squadron which might be prepared to use me on operations, and that my crew shared my frustration. Ted – a silvery-haired, fatherly figure – listened patiently to my complaint and suggested I leave the matter with him for a day or two before putting anything on paper. A few days later, on 4th September, I flew as a second pilot on a twelve-hour mission to bomb the Burma-Siam railway.

This first operation of mine was not especially challenging or exciting, but at least one of our crew had made a start. My main impression of the flight was the sharp contrast between the long flog out and back across the Bay of Bengal, and the ten minutes or so of exhilaration in the target area as we delivered our attack from a few hundred feet.

It also occurred to me, as we flew over the target area, that had we been shot down at that juncture – separated as we were from the nearest friendly troops by a mixture of dense jungle, hills and broad rivers extending over a distance of some eight or nine hundred miles – our chances of regaining our own lines would have been virtually nil. To take but one practical problem: even had we managed to evade capture by the enemy and survived the leeches, the snakes, the crocodiles and the malaria, my army marching boots, handsome and well soled as they were, were hardly likely to have lasted the long trek to our lines. And in retrospect, I know that we would not have been much better off had we been obliged to ditch at sea, unless we had come down on the last stages of the return flight – say within a hundred miles of the Sunderbans – such was the limited extent of our air/sea rescue coverage at that time. But with more immediate matters to occupy our minds, and being blessed with the unquestioning optimism of young men, I doubt if it would have cost us much sleep even had we known the worst.

Enemy Reaction Apart
Enemy reaction apart, there were certain factors fundamental to our operations – not least, our aircraft.

Entering the cockpit of a Liberator after the Wellington 1c, was like stepping up from a Land-Rover to a Jaguar. Other than the throttles, which were conventional in design, spring-loaded toggle switches replaced cumbersome levers; the seats were instantly adjustable in height, reach and rake; the cockpit lighting was precisely located and infinitely variable; the propellers could be fully feathered (to reduce drag in the event of engine failure); there was an automatic pilot, a superb one at that; and the criterion in laying out the cockpit seemed to be – will it suit the pilot? What a weak-kneed bunch those Liberator designers must have been, allowing the chief test pilot so much say in the matter! But whoever deserved the credit, the result was a cockpit splendidly matched to long sorties: my own operational sorties were to average over ten hours – with some over fourteen – so I speak from the heart.

To achieve sorties of that duration, where we were attacking targets between 800 and 900 miles from our base – occasionally over 1,100 miles – the aircraft was capable of carrying huge quantities of fuel: over three thousand gallons if necessary. Attacking the more distant targets however, was not achieved without penalty. Bomb load had to be sacrificed. First, it was not possible to carry the fuel necessary without usurping part of the bomb bay. And second, the weight of the extra fuel had to be offset by a reduced bomb load if the aircraft's all-up-weight was to be kept within the take-off limit. On short flights of six hours or so, we had no such problems and could deliver up to ten thousand pounds of bombs, with every foot of the bomb bay crammed with high explosive or incendiary.

No aircraft design is perfect: compromises have to be made. Perhaps the Liberator's most critical feature was its fore-and-aft stability during take-off and the initial climb away. Operating – as we normally were – at or near the maximum all-up-weight, it was important to keep the centre of gravity within carefully defined limits. And at our pre-flight briefings we were told exactly how our crew members were to be distributed within the aircraft until we were safely airborne. Given that we complied, there was no problem.

For its defence, the aircraft fairly bristled with gun turrets – nose, mid-upper, tail and belly: the latter a retractable ball-shaped turret. In addition, a free-mounted gun was positioned at both beam hatches. All guns were of half-inch calibre, and each turret had a pair of guns. On flights of up to ten hours, where the penetration, attack and withdrawal phases took place in daylight, every gun position was manned. But attacking the more distant targets we operated at night and dispensed with the front and belly gunners. Neither was there any requirement to man the beam gun positions, so we were able to enhance our fuel/bomb load still further by leaving behind the second wireless operator/beam gunner. (The full crew comprising eleven persons – two pilots, flight engineer, navigator, two wireless operators, airbomber and four gunners.)

The thrust required to propel this impressive inventory of fuel, bombs, men and equipment, was provided by four Pratt and Witney radial engines (Wasp R1830). Flying in formation, the way one handled the throttles was dictated largely by the leader's tactics. But flying singly – and once one had mastered the knack of putting the Liberator into the optimum cruising attitude, known at the time as putting the aircraft 'on the step' – those engines could be coaxed into quite miserly rates of fuel consumption: on the longer raids we averaged well over one air nautical mile per gallon. Such economy, of course, demanded appropriate engine handling: essentially, operating in the lean mixture range with a combination of relatively high boost and low rpm. This meant that we went nowhere in a hurry – typically 158 m.p.h. (IAS) outbound and 155 m.p.h. inbound: hence the lengthy sorties. But the sheer reliability of those engines gave us great peace of mind, especially on the longer flights and in bad weather, and if one engine should malfunction, how comforting it was to have seventy-five per cent power still in hand.

And now the bad news: the Liberator's flying controls. If my shoulders are well developed, it is in no small measure thanks to the Liberator. Applying elevator and aileron was very much a two-handed operation, often calling for the use of arms and shoulders as well as hands and wrists: a particular bind flying in close formation. Given the length of our formation sorties, the USAAF Liberator squadrons operating alongside us recognised this by introducing a miniature, fingertip control column which signalled the control requirements through the auto-pilot. This device, which removed the sweated labour, would have been a boon during defensive formation flying and pattern-bombing runs too. But during my time certainly, it did not come our way. Viewed from the cockpit, the other bugbear was the occasional over-speeding propeller during take-off, which now and then gave moments of anxiety; and on one occasion led to the loss of one of our crews.

Also fundamental to our operations was the weather. In general, in that region it is dictated by two air streams: the North East Monsoon, which blows from October to May/June and brings dry, bright conditions; and the South West Monsoon, which blows for the remaining four to five months of the year and brings (for the aviator at least) severe turbulence, lightning, heavy rain, icing, and cloud extending from a few hundred feet above ground level to thirty or forty thousand feet and beyond. (It is precisely because the North East Monsoon is by comparison such a non-event, that when people in India speak of 'the monsoon' they are referring to the one from the south west.)

The boundary separating those air streams, the Inter-Tropical Front (ITF), boasts the worst of the weather. For four months of the year, that front usually lay between our airfield and the target, and had to be negotiated out and back. Daytime conditions, with the turbulence at its peak, were the worst: having no on-board radar, we could only grope our way through. Crossing the ITF at night was less difficult: apart from the reduction in turbulence from surface heating, the core of each thunder cloud was illuminated by lightning at frequent intervals, which helped considerably in threading our way through.

There was no panacea for negotiating the ITF in daylight. Some methods worked, and some did not. But they had things in common – the experience was never less than interesting; occasionally it was frightening, and once in a while it was lethal. Only a month before

I joined 356 Squadron three Spitfires had been destroyed in a thunderstorm while flying from Imphal to Calcutta. And of the three aircraft which survived, I believe that only one actually made it to Calcutta; the others being obliged to make emergency landings short of their destination. Another account of the weather in the Imphal area comes from Group Captain 'Ace' Newman, who recently described to me how his Blenheim was turned on its back in a cumulo-nimbus cloud.

Even an aircraft the size of a Liberator could be half taken over by the sheer brute force of such thunderstorms. On one occasion Stringer-Jones, flying just beneath the ITF over the Bay of Bengal, was finally obliged to put his crew into ditching positions as he was forced lower and lower by the strength of the down-draft, only the cushioning effect of the sea – reacting to the down-draft – ultimately saving the situation. From Stringer's account, I can assure my reader that his holiness Pope John was not the first person to kiss the ground on leaving his aircraft.

On solo missions, some pilots would make quite a wide detour in the hope of finding a gap in the ITF, rather than attempting to go through it. And it was by no means unheard of for missions to be abandoned in this way through shortage of fuel. But most of us, if the alternative appeared to be a wide diversion, had a shot at going through. Before entry (if there was any choice – *below* the icing level), we disengaged the auto-pilot, tightened our safety harnesses, closed the windows and hatches, turned the cockpit lighting on full and wound in the trailing aerial. Once in cloud, we altered course as necessary to avoid the really ugly, black-as-pitch sectors but otherwise bored straight ahead. As we approached the core of the ITF, the flying controls would often twitch and jerk with the unpredictability and violence of things possessed as they fed back the buffeting from the storm. But the instinctive reaction of not yielding an inch to these convulsions had to be curbed if we were to avoid yet more strain on the control linkage. Consistent with remaining more or less straight and level, Jack Clarke and I (both of us on the controls) contented ourselves with riding out the worst of these convulsions rather than attempting to resist them completely. More often than not we were through the immediate barrier of thunderstorms in a matter of minutes, though there were times when it seemed a good deal longer.

Another feature of the environment was the presence of large birds and the problem they posed if you hit them. And whilst too localised to compare with the dangers posed by the worst of the weather, as an impairment to our operations they were something we could not ignore. Varieties like the kite hawk and the vulture were always a potential hazard in daytime as we climbed away after take-off, and also when we approached to land: a similar problem sometimes confronting us over enemy territory during our low-level daylight attacks.

Hitting one of these large birds, the vulture especially, the damage to the aircraft could be quite extensive. Striking the leading edge of the Liberator's wing, for example, the weight of just a single bird could push that portion of the metal skin back some eighteen inches to the wing spar, leaving a jagged hole a foot or two across; which did nothing to improve the wing's lifting properties, or the fuel consumption for that matter as we opened the throttles to compensate for the extra drag. More worrying was the possibility of injury to the crew from a head-on strike on the cockpit or the front gun turret. In flight, we only seemed to spot these birds at the last moment, and if it looked like being a head-on collision, the drill was to pull back on the stick and hope that the bird would pass beneath, where if it did hit the aircraft, the consequences might be less critical.

As the day hotted up, with the thermals reaching upwards to many thousands of feet, so the vultures went progressively higher. Frequently they wheeled about up to five or six thousand feet over the fringes of the large cities, carving their graceful orbits for minutes on end without as much as a single beat of their wings. (What a strange contrast these most elegant of fliers present when they are huddled together on the ground or in the branches of a tree, as they flop about awkwardly, taking bad-tempered pecks at one another). Sometimes there were dozens of birds in the same thermal, in which case they were easy enough to spot. It was the lone bird, difficult to pick out, which was the real menace.

For the most part the vultures seemed to feed on dead carcasses of one sort and the other, and as such could be found over open country as well as city outskirts. In the latter case we assumed that part of their attention was devoted to the ghats where the Hindus burn their dead in the open. But even more attractive targets for their attention were the Towers of Silence, on which the naked bodies of the Parsee dead are laid out, open to the sky, for the very purpose of being disposed of by the vultures.

Whereas vultures tended to avoid human contact, the kite hawks showed no such shyness, and positively exulted in displaying their skills. They were to be found between ground level and a few hundred feet, near the mess halls on the camp and in the vicinity of villages and city hotels where they maraud for food in their dozens: one moment plummeting earthwards, wings folded, and the next soaring upwards above the rooftops, wing-tip feathers spread delicately like the fingers of a hand, their forked tails twitching, first one way and then the other, as they manoeuvred to avoid colliding with one another and to position for the next swoop on the scraps below. As an aerobatician, the kite hawk can have few rivals.

Before leaving the subject of the kite hawk, in his Indian habitat he must be one of the cheekiest birds the world over, as many a serviceman can testify. I recall helping to serve the airmen's dinners one Christmas – which entailed carrying the loaded plates a short distance in the open – when the foraging attacks from the kites were so accurate and persistent that we were quickly reduced to carrying one plate at a time, in order to have an arm free with which to ward them off. Not without good cause was the name kite hawk corrupted by the British serviceman to something coarse and disparaging.

Off the Leash
Even if I could recall the thirty-five operations my crew and I carried out with 356 Squadron, I would not dream of boring my reader by describing them individually, much of the action being of a repetitive

nature. But a detailed look at a handful of those operations will perhaps convey the flavour of our work better than stopping short at a broad description of the various types of raid (something I do later on pages 101 to 103).

Inevitably, in selecting those operations which are etched deepest in my memory, I deal mainly with those where things did not go according to plan, and I risk leaving the impression that as a crew – and to a lesser extent, as a squadron – we were frequently in crisis. Such was not the case, and let me assure my reader that the vast majority of our sorties were competently discharged and successful in their outcome.

My memory apart, the basis for each operational account is the entry made at the time in my RAF pilots' log book; augmented with information from the squadron, wing and group operations records books, made available to me by the Public Record Office.

On 24th September, I finally set out on my first operational sortie as an aircraft captain. Having already flown on two operations as a co-pilot, I was more than glad to be off the leash and flying with my own crew.

Since our arrival at Salbani we had been joined by Phil (Arthur Phillips), our ball-turret gunner, a fresh-faced young man from Fallowfield, near Manchester. Although the baby of the crew, both in appearance and age, Phil looked to have a very composed temperament, and, as he showed me over his turret, his approach was strikingly professional. It would probably take a few sorties before I really knew what made young Phil tick, but from first impressions I was more than pleased with our new acquisition. Come to that, the whole crew could expect to discover quite a bit about one another under the stress of operations in the weeks to come, given that we survived. Roy Brown too had joined us for this sortie, in the capacity of front gunner, but he did not remain with us for more than a few sorties, after which Geoff Gradwell filled that position on a permanent basis and did so with good-natured efficiency.

Our target comprised the railway repair shops and station sidings at Maymyo, thirty miles east of Mandalay. This was evidently an important link in the enemy supply chain, since we had already attacked it less than a fortnight earlier. On that occasion, as Ron Johnson's co-pilot, I had watched with satisfaction as our sticks of high explosive ran obliquely across the tracks and station buildings, leaving a trail of craters and twisted wreckage: but soon repaired it seemed.

Like the earlier Maymyo raid, this was a daylight attack: Hugo Beall's plan being to run in on the target in a single formation of twelve aircraft; comprising three boxes of four aircraft flying in close line-astern. By flying in a tight formation at half wing-span intervals and releasing our bombs simultaneously, the object was to lay a carpet of destruction tailor-made to fit the target. Never was the fitter of an Axminster or a Wilton more painstaking than Tim Adams, Hugo's air bomber, as he pored over the target map ahead of those pattern-bombing raids, calculating the requirement both in terms of the size and shape of the formation and the intervals at which the bombs should leave their racks.

That morning we rose at five-thirty, began our briefing at six-thirty, and were assembled at our aircraft shortly after seven. Being in the last box to take off, we had ample time in which to complete our pre-start checks. I had slept well; obeyed the no-drink rule the evening before; had a good breakfast; and felt generally relaxed as I turned the propellers by hand and chatted with Sergeant Todbury and other members of our ground crew. Meanwhile the rest of my air crew were completing their individual pre-start routines. All checks completed, we heard other aircraft start up in adjacent dispersals as we waited quietly for our turn: each of us with his private thoughts.

We started up on time; the engine checks presented no problems, and we took our allotted place in the queue of aircraft waiting for take off. Just then Hugo lined up, opened his throttles and accelerated on his take-off run. He was airborne comfortably before the end of the runway, by which time his number two had already started his take-off; and so it progressed, until we too were under full power and gathering speed down the runway.

The best method of getting a Liberator into the air with a full load was still in dispute among the pilots at that early stage in the squadron's life. There were those who advocated using every yard of runway, regardless of the circumstances; if necessary forcibly holding the nose down until the end of the runway in order to gain every bit of speed possible before lift off. But I was persuaded by the arguments of the more scientific school of thought: namely for getting the nosewheel a foot off the ground as soon as the elevators became effective, and just holding it in that position – if the principles of flight were worth anything, take-off would follow automatically at the right moment, surely!

But to return to the reality of the moment. This was my first full-load take-off at the controls, and there was no doubt it felt quite different from usual. We seemed to be accelerating desperately slowly, and when the elevators became effective and it was time for me to raise the nose, I felt the gut temptation to follow the extreme tactics of the 'hold it down' school. But reason prevailed, and using my normal technique we unstuck with runway to spare. Nevertheless, as I settled quickly into a climbing turn to link up with our box leader, I had a distinct feeling of relief that science had not let me down.

As we slotted into position on our box leader's left, we were heading almost due east. At that stage, still west of Calcutta, we fanned out into loose formation; and with the early morning sun sitting just above the horizon and on a south-easterly bearing, I moved forward to keep station abeam of our leader in order to avoid its glare.

Abreast of Calcutta, as I glanced beyond the box leader, I caught a glimpse of the Sunderbans (mangrove-covered islets), as ever guarding the mouths of the Ganges. We might be off on our first raid as a crew, but to such eternal features of the landscape it was just another day. The sun had not yet created thermals at our height, and although I could see a build up of cloud ahead, the air was agreeably smooth as we headed across the Bay of Bengal towards the Arakan coast. We were holding good station, and at that stage I was having to work just hard enough to find the task nicely absorbing.

Chittagong, on the Arakan coast, was cloud-free and passed almost directly under the nose: nicely fixing our position before the poorly featured terrain of western Burma. Over the early ridges of the Chin Hills, before entering enemy airspace, we tested our guns and our box closed up into defensive formation. The mood within the aircraft too had tightened: the intercom silent except for essentials. At this stage we had been flying for two hours and were a little over half way to the target.

In the next hour and a half, I found the workload more than nicely absorbing. As the sun climbed, so the cockpit temperature and the turbulence increased. And although we managed to stay clear of cloud, it was only at the expense of frequent alterations to our course.

No enemy fighters challenged us as we neared Mandalay, but that did not lessen the thought that they might do so at any moment. Somehow our intrusion appeared all too easy, and the tension which had settled on us as we entered enemy airspace remained to keep us alert.

As Hugo edged us past Mandalay and we headed towards the higher ground containing our target, so the cloud cover beneath us thickened, and it was already obvious that Maymyo might be obscured. However, the secondary target at Myitnge lay further south, and the cloud cover in that direction was well broken at that stage. Whatever happened, we would have something to attack.

That was the position as we headed towards Maymyo. Shortly afterwards normality was rudely shattered. We had moved in to half wing-span lateral spacing and were on the bombing run itself, with only a minute or so to bomb release, when we entered solid cloud. For a few moments I held on to our box leader. Then, quite abruptly, I lost him. I altered course twenty degrees away from the formation; held that heading for fifteen seconds, then resumed course. I also called to Jack for climbing power and set about extricating ourselves from the cloud: the hills rose sharply beyond Maymyo, and I had no wish to cut short so many promising careers.

Emerging from the cloud tops, I expected to see the rest of our formation in a loose gaggle to my right, but I could see only a solitary Liberator about two miles away, already heading in the general direction of the secondary target. We followed. Our instructions were to act independently if the formation broke up, and I could see no future in making a solo run on Maymyo. Hugo Beall was a man of some determination, and since he had elected not to chance a run below cloud, I reckoned it was unsafe to do so.

As we neared the secondary target – the railway workshops and station sidings at Myintge – the Liberator we had seen earlier was already on its bombing run. Black puffs of smoke were breaking out all round it as it received the attention of the anti-aircraft defences. A few minutes later we received similar treatment as we ran in, but sustained nothing worse than a minor hit on the front turret. Thankfully, Japanese technology in anti-aircraft shells seemed to be lacking, because there appeared to be little wrong with their aim: not if one could judge from the proximity of those black puffs. Our bombing run was a classic, with no alterations of course in the final seconds. And I was just indulging in a little mental self congratulation at not being put off by the gunfire, when the words 'hang up' came over the intercom. It was not our day.

I turned in a wide arc to make a second run, only to find the rest of our squadron queuing up to run in. It would be several minutes before we could make a second run. At this point Red voiced concern about our fuel, and since the secondary target was not of vital importance, I agreed to Jock's suggestion that we turn for home and attack the Sagaing–Monywa railway line en route.

Turning on to our new heading, I engaged the automatic pilot. As a rule it behaved superbly but, on this occasion, it stubbornly refused to hold a steady course. Having engaged it several times with the same irritating result, I switched it off and invited Jack to take over the controls. By then I had been at them for four and a half hours non-stop, and Jack was no doubt as much in need of activity as I was in need of a rest.

To avoid the risk of another hang up, we modified our drill for attacking the railway line, whereby, having pressed the bomb release switch, Jock would immediately operate the bomb jettison lever. This time it worked without a hitch and, more important, we scored an almost perfect hit: our stick of bombs straddling the railway line at a shallow angle and severing it at two points. We would, after all, have something to show for our efforts.

All we had to do now was simply to get home. As we turned on course for base, Jock's estimate was that it would take us in the region of three hours. In the event it was to take days rather than hours.

Proceeding westwards on our return flight, the cloud cover beneath us rapidly became complete, so that we could no longer map read. And with the sun almost directly overhead, a check on our course using the astro compass was also out of the question. To add to the growing list of problems, Terry announced that his direction-finding equipment was unserviceable. However, as I acknowledged Terry's news, I was not really concerned. The cloud cover would almost certainly break up over the Arakan coast, where we could hardly fail to pinpoint our position.

Fifteen minutes beyond our estimated time of arrival (ETA) at the Arakan coast, the cloud cover beneath us was still solid. I carefully resynchronised the master compass repeater in my cockpit and received confirmation from Jock that I should maintain our westerly heading. I was still pondering the unusual feature of solid cloud over the coastline, when a few miles ahead a gap appeared in the cloud cover. Shortly, there would be no question as to our whereabouts.

It was scarcely credible. Through the break in the clouds, lay not the Sundarbans or Calcutta, but a ridge of towering hills! Initially, Jock too was completely baffled, but recovered himself to ask that I check the aircraft heading on my emergency compass: a tiny, primitive instrument for use in the unlikely event of the master compass becoming unserviceable. It showed us heading due north! And since the only hills of those proportions were well to the north of our intended track, that tatty little emergency device was clearly correct, and the master compass repeater – that sophisticated charmer occupying the central position on my flight panel – was reading some 90 degrees in error. (Perhaps the failure of the auto-pilot to hold a steady heading could now be accounted for.)

Calling for a visual pinpoint, I was told we were not carrying the map for that area. In exasperation, I remonstrated with Jock for not complying with the standard practice of carrying the map sheet adjoining our intended track. He barely heard me out, replying with hurt pride that the one containing our track already took in the area for a hundred miles north of it. He had a point perhaps, but the fact remained we were lost, and we were without the one means of establishing our position and heading for an airfield. As if I could not see it coming, Red Rigby suddenly tapped me on the shoulder and whispered, 'Forty-five minutes fuel remaining, Skipper.'

Nothing concentrates the mind like an emergency; a situation we were fast approaching. With Jock I conceived a rough plan of action, our tempers having subsided as quickly as they had risen. We guessed the high ground to be the foothills of Shillong and concluded that by heading south west, in the general direction of Calcutta, we stood a reasonable chance of finding an airfield. I announced the plan to the crew, and said that if we had not found an airfield by the time our fuel reserves were down to fifteen minutes, I would have them bale out and either follow them or attempt to crash land. I had everyone acknowledge.

The next thirty minutes seemed savagely compressed. Within the first ten minutes we spotted a large clearing carved out of bamboo jungle. It was a curious feature, and my guess – although it hardly made sense in that isolated area – was that it was a railway marshalling yard under construction. Otherwise, why the railway lines down one side and the concrete platforms beginning to sprout from one end of the site? The whole clearing, however, was vaguely the size of a runway, and if it was ultimately intended to bear the weight of a train, or even an engine come to that, I reckoned it should stand the weight of a practically empty Liberator. The problem was that the usable surface appeared to be raw earth, rolled perhaps, but otherwise untreated. And there were odd pieces of earth-moving equipment dotted about. I noted this 'marshalling yard' as a last ditch possibility and we proceeded with the search.

Ten minutes later we had found nothing better, let alone a formal airfield. Meantime, I had Jack check Red's calculations on fuel remaining. Red had not exaggerated, I would have to put the aircraft

ABOVE. My Liberator aircrew *left to right,* **back:** Ted, Smithy, Terry, Roy, Red. *middle* Jack, Author, Jock. *front:* Chile, Phil. BELOW. My Liberator ground crew. *Sergeant Todbury at centre front.*

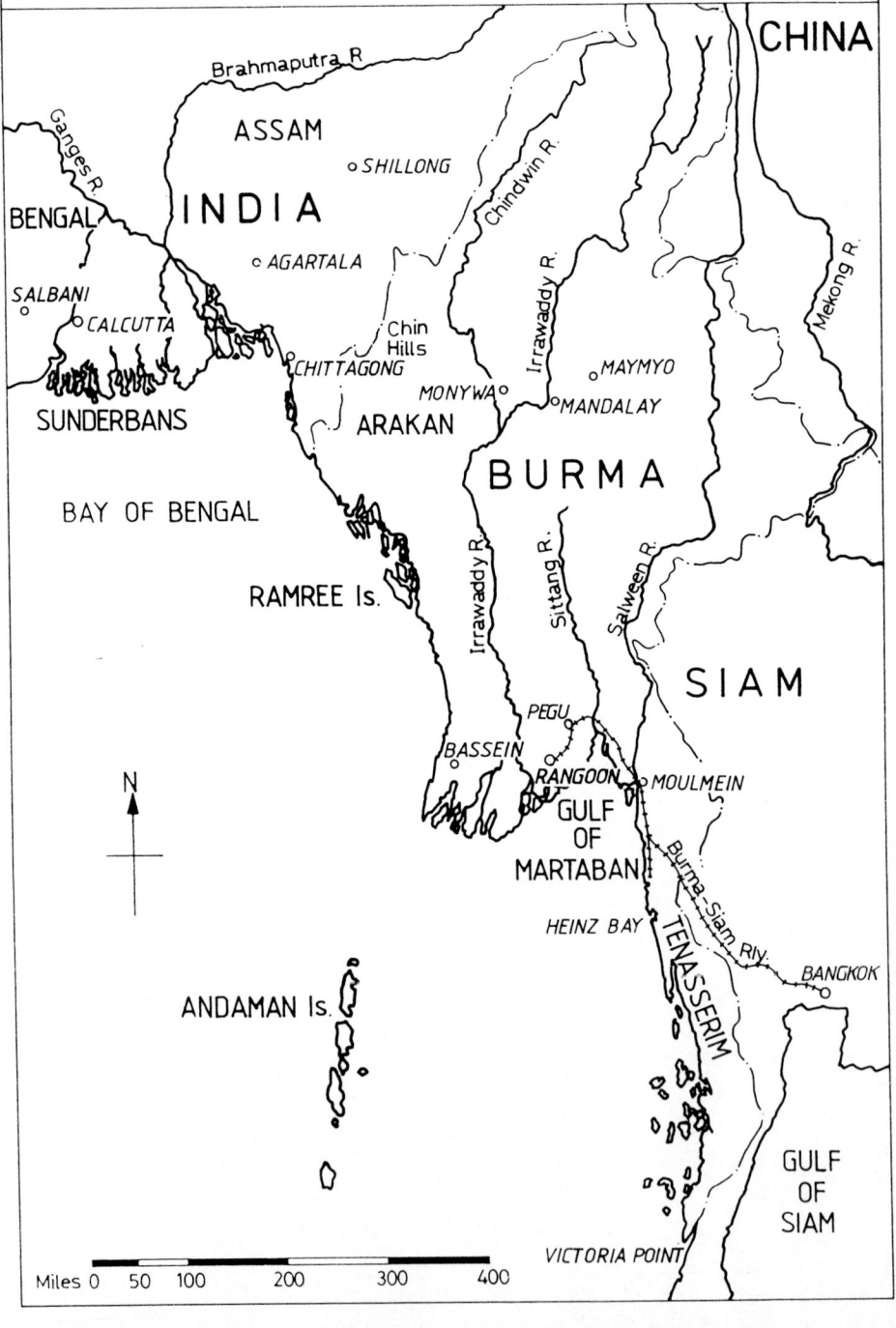

STRATEGIC AIR FORCE (SAF)
GENERAL AREA OF OPERATIONS 1944-45

CHINA

Brahmaputra R.

Ganges R.

ASSAM

o SHILLONG

BENGAL

INDIA

Chindwin R.

Irrawaddy R.

Mekong R.

SALBANI
o

o CALCUTTA

o AGARTALA

Chin
Hills

CHITTAGONG

MAYMYO o

SUNDERBANS

MONYWA

o MANDALAY

ARAKAN

BURMA

BAY OF BENGAL

Irrawaddy R.

Sittang R.

Salween R.

RAMREE Is.

SIAM

N

PEGU

BASSEIN

RANGOON o MOULMEIN

GULF
OF
MARTABAN

Burma-Siam Rly.

HEINZ BAY

TENASSERIM

BANGKOK
o

ANDAMAN Is.

GULF
OF
SIAM

Miles 0   50 100   200   300   400

VICTORIA POINT

356 Squadron Liberator

356 Squadron

ABOVE LEFT. Pattern bombing of artillery position at Ywabo Hill, February 1945, *target outlined.* ABOVE RIGHT. Precision attack on Lewe airfield. BELOW LEFT. Attack on Burma-Siam railway locomotives and rolling stock, November 1944. BELOW RIGHT. Makeson railway workshops, Bankok. 'defences gave us the mixture as before'.

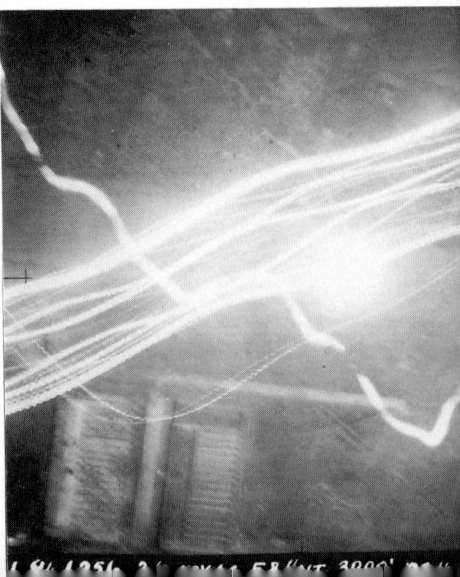

Destruction of a Burma-Siam railway bridge. ABOVE. The attack. BELOW. The result – (*Photo's courtesy of Imperial War Museum*).

The Shwedagon Pagoda, Rangoon.

down or abandon it in the next twenty-five minutes. I saw our options clearly: either we could press on blindly, hoping an airfield would turn up within the next ten minutes, or I could turn back now and have time in hand setting myself up for a landing on the jungle clearing, alias 'marshalling yard'. I opted for the latter, warts and all. On the way back, I asked each crew member if he would prefer to bale out, since the landing would be a bit chancy. To a man they preferred to stay abroad.

We sized up the clearing in a low fly past. And although it showed signs of being waterlogged at the touch down end, it looked firm enough from a bit further in to attempt a wheels-down landing. If my judgement was right and we landed with the undercarriage down, we stood a chance of ending up with a completely undamaged aircraft. Alternatively, if I had misread the situation and the ground was too soft, we would incur considerably more damage than in a belly landing, and more than likely sustain injuries to the crew. I decided to land wheels down, emphasising the need for everyone to brace themselves before touchdown, and to remain braced through-out the landing run. With the small amount of fuel remaining, the risk of a serious fire would at least be reduced should the fuel system be ruptured.

On the downwind leg, Jack suggested I fasten my shoulder straps. I had secured my waist belt, which was all we normally used for landing, and the shoulder harness had not even crossed my mind. Jack was right, of course, given the circumstances. I grunted my thanks and gave him control, as I followed his suggestion. Pinned firmly to my seat by the full safety harness, as we turned on to the approach I felt more like a chicken trussed up for the spit, and I remember hoping the analogy would be taken no further.

A more immediate problem intruded. Although Terry had fired off numerous red Very cartridges to show that we were in serious trouble – and our lowered undercarriage should have indicated our intention to land – our landing path was still obstructed by hundreds of native labourers. I reckoned we had enough fuel for one more overshoot and this time I flew very low, just above their heads, which at least scattered some of them to the verges.

Four minutes later we came in on what had to be our final approach, with Terry now blasting off our remaining Very cartridges regardless of their colour. As I motored in just above the stall and levelled off for touchdown, so the remaining squatters surged to the sides of the strip like so much flotsam. We were down and the undercarriage legs were holding. Suddenly an obstacle I had not picked out loomed up, threatening to impact with our starboard outer engine. A touch of rudder and it was avoided. We were slowing down rapidly and, with Jack's assistance, I hauled back on the control column to keep the nosewheel from digging in. In the final hundred yards of our run, I could almost feel the crew willing the undercarriage not to fail, as the nosewheel began to sink into the soft ground.

We had stopped: distinctly nose down, but the undercarriage had held; and more important, the crew was unhurt. Jack and I exchanged a broad grin, and I shouted to everyone to leave the aircraft from the beam hatches in case the undercarriage should finally collapse.

As we finished switching off the cockpit equipment and I was about to close my side window, a raucous American voice called up from below, 'Say captain, how come you landed here? There's a perfectly good airfield at Sylhet, only fifty miles north.' Clearly he felt we were cluttering up his strip. I gave him a short, sharp reply and banged the window shut.

By the time I reached the brash owner of that voice at ground level, he had been joined by the real boss of the site, a British major of the Royal Engineers. The latter was a man with a smile, a welcoming handshake, and a real concern to know what help he could give us. What a difference in their attitudes and in my response. A moment ago my hackles were up, and I had felt like gripping the American by the throat. Now I was responding affably to the Major's invitation to join him in the Mess, and to discuss our problems over a cup of tea.

As I had stepped out of the aircraft, Ted came forward, shaking me warmly by the hand and muttering his thanks for a safe landing. Poor chap, he must have feared the worst and his sense of relief had quite overtaken him. I hoped the British army would not think the

Royal Air Force was normally quite so theatrical. On reflection, perhaps Ted, being the old man of the crew, sensed better than most of us how lucky we'd been to find somewhere to put down.

We had landed at Shamshernagar, an airstrip in the early stages of construction on the Assam border. The railway lines down one side were there for the delivery of materials, and the embryo concrete platforms were the start of the runway itself. The reluctance of the labour force to evacuate the strip, in spite of our signals, was simply explained. Each week a Dakota from Sylhet not only dropped the mail in a low fly-past, but usually beat up the place just for the hell of it. Until our final approach, we were seen as a different aircraft, with an even more exhibitionist crew, but about to perform the same task.

In the Mess that night, the Royal Engineers could not have been more hospitable; producing a bottle of whisky – their ration for the month – and insisting that we help them kill it to celebrate the opening of their airstrip. It was an evening of great bonhomie. Meantime, we sent a signal to Salbani advising our condition and whereabouts; removed certain sensitive and attractive items from the aircraft; and arranged for the aircraft itself to be properly guarded. As I lay in bed that night pondering the events of the day, my overriding impression was of the lucky twist at the end. Before drifting into a fitful sleep, I whispered a small prayer of thanks.

It was forty-eight hours before we were able to make two-way contact with Salbani and to complete arrangements for our recovery to base. There was no question of being able to fly our aircraft out of Shamshernagar at that stage. The simple fact was that the surface of the strip, raw earth as it was, was too soft to take the weight of even a modestly refuelled Liberator and at the same time give it anything like an acceptable chance of accelerating to take-off speed. The day after our landing, the main wheels had already sunk a further three to four inches, even with empty fuel tanks. Before our aircraft could be recovered – short of dismantling it – the runway would have to be completed, and that was obviously going to take some months.

Contact with our base established, we said goodbye to our army friends and went south by rail to Agartala. By European standards, we must have looked an odd bunch, dressed for flight and weighed down with the breech blocks from our machine-guns, our IFF (Identification, friend or foe) set and our parachutes. But this was India, where by our standards, the exotic is the norm. The crowds at the railway station and our fellow passengers hardly gave us a second glance.

Squadron Leader 'War' Harris, our squadron's senior flight commander, flew to Agartala on the morning of 27th September to collect us, and we were back at Salbani in time for lunch. Before landing at Agartala, 'War' had flown over the Shamshernagar strip to assess the problem of recovering our aircraft. And he too was in no doubt that the runway would have to be completed before it could be flown out. He was also kind enough to compliment me on getting the aircraft down without damaging it. Since he appeared a bit stuffy as a rule, I found his friendly approach on that occasion quite reassuring. Perhaps I should explain that he acquired his rather striking nickname simply because his first three initials were WAR.

Everyone at Salbani seemed extremely pleased to see us. After twenty-four hours without word, we had been posted missing, and the popular belief was that we had probably flown into high ground while attempting a run below cloud. That evening I resumed my seat in the Mess poker school, taking up where I had left off four evenings before. The only difference being that I had to take some good-natured banter about the misfortunes of people who are lucky at cards.

The following day, the adjutant asked me discreetly if I would like to take my crew to Calcutta for a forty-eight-hour break, before we resumed our flying duties. He was acting on instructions from Hugo Beall, and although I brushed aside the offer on the grounds that we did not need the break, it was a thoughtful gesture. On walking into the Mess the day before, Hugo and I were quickly into a game of shove-halfpenny, and as I went to accept a cigarette from him, my hand had trembled alarmingly. So much so, that Hugo, recognising my embarrassment, had put the tin down and left me to help myself. Perhaps it was that display of tension which had triggered the adjutant's question.

Glad as everyone was to have us back, there was the need to account to our group headquarters for the temporary loss of an aircraft. The official enquiry found that we had been negligent in not cross-checking the reading of the master compass with the emergency compass as a matter of routine. But the enquiry also revealed an extraordinary connection between our navigation error and that of Hugo and his crew when he had led the squadron progressively further north of the intended track when returning from one of the squadron's working-up exercises (see page 76).

Following Hugo's incident, the master compass in his aircraft was found to have a progressive error and was replaced with a new unit: the faulty one being returned to the instrument section for rectification and in due course installed in my aircraft. In short, better men had been led astray, if not to the same extent.

All this took time to come out, and meantime we had resumed operations as though nothing had happened, except that for Jock and me the emergency compass took on a new and lasting significance. A month or so later, when the enquiry was complete, I was officially reproved in my capacity as the aircraft captain for failing to ensure full compliance with navigation procedures laid down in Air Staff Instructions: a voluminous and rather dry tome to be found in every squadron crewroom. But given the circumstances and the fact that by then we had a number of successful operations to our credit, my reproof was delivered almost apologetically.

Curiously, as a crew we gained from the Shamshernagar affair. Although the mess we got into was ultimately of our own making, we did extricate ourselves successfully: albeit with a generous slice of luck. And having risen to the challenge and got away with it, the effect was to weld us together much quicker than if our early operations had been more routine.

Some months later I was urged to make my way to the airfield where, I was told, an old friend of mine was passing through. It was the aircraft we had put down at Shamshernagar: now resplendent with new bomb doors, the original ones having buckled beyond repair

as the aircraft had sunk lower into the soft earth. On completion of
the runway, the aircraft had been flown out to continue a useful life
at another unit. But not, as I had been told at the time of my reproof,
before a maintenance party had replaced a host of miscellaneous
items removed by souvenir hunters, in spite of the guard I had
arranged. Well, things might have been worse, and perhaps there is
an otherwise modest Indian dwelling which still boasts a handsome
aircraft clock.

Life at Salbani

Although we shared the Salbani base with 355 Squadron, the two
squadrons saw little of one another on the ground, except at joint
briefings and occasionally at the station cinema. The fact was that
the two sets of living quarters were on opposite sides of a large
airfield, where the by-word during construction had been *dispersal*.

Our domestic accommodation was located among trees and well
established thickets of bamboo, about three miles from the airfield
by road. The officers' sleeping quarters were separate from the Mess
itself and consisted of eight or ten bungalows: each with a verandah
front and back, and most comprising five single rooms. The function
of the verandahs was to keep the rooms cool during the hot season
by shielding the walls from the direct rays of the sun, but they were
useful too in other ways: in particular the rear verandah, which was
used for our ablutions. The high point of the latter was our daily
bath, which we took squatting cross-legged in a shallow canvas
contraption: an item on permanent loan to each officer, along with
his canvas washbowl, his collapsible camp-bed and his mosquito net.

We took our main meals in the Mess where, provided he passed
muster with the head waiter, one's personal bearer (servant) took his
turn waiting at table, smartly turned out in a white uniform, the
RAF colours displayed on his turban and on a broad belt around
his waist. But afternoon tea – as well as water for drinking and
ablutions – was brought to our bungalows by the bearers.

One's bearer really was a man of some importance to the daily
routine. It began with him gripping one's shoulder through the

mosquito net and bringing one to consciousness with the gentlest of shakes, and a soft '*Sahib, sahib, saht budgie sahib – char* bringing *sahib.*' If that did not work, after a decent interval – sometimes it seemed less than decent – the shaking became more persistent and the announcement that it was seven o'clock and that he had brought the early morning tea, became that much more urgent and emphatic. The morning call was perhaps the bearer's most delicate task, since if his message got through it was hardly likely to gladden his master's heart, and if he let him oversleep, he stood to receive the rough edge of his tongue.

By the time one had swigged the tea and decided to face the day, hot shaving water had appeared in the canvas washbowl on the back verandah and the bearer was squatting nearby polishing one's *chaplis* (sandals). Returning to the bedroom, with ablutions completed, more magic – fresh khaki shorts, shirt and stockings were already laid out on the bed and one was being asked what *dhobi* (laundry) was to be done. Having received his orders for the day, and assuming he was up to standard, Abdul, Wallaya or Richard – whatever his name might be – would dash ahead to the Mess to wait on the breakfast table.

It was generally accepted that Punjabis and Kashmiris made the best bearers, provided the right calibre of man was available. For the most part men from these districts were warrior-like figures with proud faces and straight backs. Frequently they carried impressive references from British military personnel and members of the Indian Civil Service. By the time I joined the squadron, however, Punjabis and Kashmiris out of the top drawer were no longer available, and I had no alternative but to recruit from the local Bengali community.

That is how I came by Wallaya, a '*jungli wallah*' ('man from the jungle' – a term of derision used by the Punjabi bearers) if ever there was one. But a good-humoured '*jungli wallah*' nevertheless who, although he lacked proper training, greeted every instruction with a broad grin and did his best to give satisfaction. In addition to his good humour, Wallaya had two more endearing qualities – first, he was never rattled under pressure, and second he was extremely appreciative of any little gift, no matter how small. And here I am not talking about gifts in lieu of wages: he received a monthly wage from me as a matter of course.

But one day I learned that even Wallaya had his standards in the matter of gifts. Whitehall had instituted a free issue of cigarettes to our forces in India – about twenty a month – which went under the brand name of 'V for Victory' or something close to it. To give you an idea of their quality: they were so dry that if you tapped them at all vigorously – to get rid of the loose bits – you were as likely as not to be left with the hollow tube. They smoked to a comparable standard. Finding them pretty foul, I offered the balance of the first issue to Wallaya, which he accepted with his customary grin, and some embarrassment at my generosity. The following month, when I went to repeat the act, I got an even broader grin, accompanied on this occasion by a polite but firm 'Nay, sahib.' Bearing in mind that he was even denying himself the opportunity of passing them straight on to a fellow villager, it was indeed damning evidence as to the standard of this Whitehall gift. It came as no surprise to learn that their wretched quality had been the subject of questions in the House, and that future issues had been suspended.

Remaining healthy in India demanded certain elementary precautions: we slept the year round under mosquito nets; shook out our boots before putting them on, just in case a small snake or a scorpion had nestled inside overnight; and when the rainy season hit us, we kept an even keener eye on the dark corners of our rooms, as the snakes came slithering out of their flooded holes in search of a dry home.

Other animal life also invaded our bungalows: some amusing or irritating, depending on one's mood, and some positively useful and welcome. In the first category came the tree rats, which used the piece of hessian serving as a false ceiling, as a kind of trampoline – holding their sporting fixtures to coincide with our afternoon siesta. Our useful visitors, the leathery-skinned geckos (lizards), with their bulging eyes and adhesive toes, came on the scene chiefly at night; when they moved across the wall with lightning speed to gobble moths and other insects rash enough to settle in the circle of light from one's table lamp.

During the hot season, with the temperature well into the hundreds fahrenheit, and the rays of the sun beating down with an almost physical intensity, even our thick-walled bungalows became oppressively hot during the day. And although latterly we were equipped with electric table fans, in my early days in India we relied entirely on the punka wallah to produce something vaguely resembling a breeze. He sat outside, at the shady end of the bungalow, tugging on a rope connected to a series of wooden battens suspended from the rafters. One of these battens – each with a broad strip of material hanging from it – straddled each room, and as long as the punka wallah stayed awake and kept up a rhythmic pulling action, the air was at least kept moving. And if he dozed off, it was not for long, as irritated shouts of 'Punka Wallah!' brought him to.

In case it appears that we led a slothful existence, perhaps I should explain something of our working routine. During the hot season, we worked a single-shift day, starting at six in the morning, and going straight through to one o'clock midday. This routine mirrored to a large extent what the local population did and, by definition, their methods produced survivors. When we were engaged in operations, siestas naturally went by the board for all concerned, hot season or not. And during the winter months, from November to March, our working day was much the same as at home.

From April through to September, the Mess dining tables carried bowls of salt tablets, and it undoubtedly helped to ward off fatigue if one took a tablet or two after sweating a lot. But not everyone found it easy to keep them down if they were taken early in the meal, and new arrivals especially were prone to get it wrong – sometimes making a dash for the door, even before they had finished their soup.

Just as it was important to replace body salt, it was necessary in the hot season also to avoid dehydration. Something the char wallah, strategically placed alongside the flight office, was more than happy to ensure during the working day, on the payment of a few annas. On the domestic site, the Mess, of course, had a plentiful supply of safe drinking water. And to store water in our rooms most of us invested in a *chattie*: a porous earthenware container obtained from the local village which kept its contents beautifully cool, purely by evaporation.

Despite these elementary steps to preserve health and efficiency, the hot season was a debilitating affair. Septic prickly heat was common among air gunners and armourers: the gunners, because in servicing their turrets, it was equivalent to working in a badly ventilated greenhouse of incredibly small dimensions; and the armourers, because not only was their job of winching the bombs into position (in a very confined area) a heavy manual one, but it frequently had to be done in the heat of the day, as they responded to last-minute changes in the bomb load. In fact the conditions under which ground crew of all trades did their work at the aircraft dispersal pans meant that none of them could expect to escape scot-free from this irritating complaint. I have put the spotlight on the armourers only because they were at special risk. With most people, once their prickly heat had turned septic, it took a decent spell of leave at a hill station to clear it up.

The high temperatures and humidity of the hot season also posed a problem in selecting the right clothing for wear in the cockpit; at least until we had climbed to three thousand feet or so. A problem which Jack Clarke and I overcame by stripping to the waist before take-off and taking it in turn to towel-down and dress in stages on the climb; ultimately arriving at our cruising altitude of eight thousand feet wearing battle dress tops and silk scarves. No doubt present-day RAF aviation medicos reading this would have kittens thinking of the flash burns we were exposing ourselves to, had we crashed on take-off. But in those days we were young and foolish; the station medical officer was an understanding chap; and the experts were not yet breathing down our necks.

I can imagine the reaction of former members of the Fourteenth Army should they read my account of our domestic life at Salbani. 'What,' they'd say, 'sleeping between sheets under mosquito nets; punka wallahs; bearers bringing morning tea: some war, eh!' Indeed, they would be right to contrast it with their own squalid conditions in the field. But, being of necessity based far behind the front line, there would have been no merit in us airmen pigging it for its own sake, and we too were stretched in doing our job, but in a different way.

Types of Raid

We attacked mostly in daylight; partly because many targets were too ill-defined to locate at night, and because in daylight we could do so in formation. The latter technique (sometimes referred to as 'pattern' or 'carpet' bombing), in which the aircraft released their loads simultaneously, produced a devastating effect, with the sticks of bombs unrolling carpet-like to cover the whole target area in a matter of seconds. Generally, this technique made up for minor inaccuracies in aim, and the very concentration of the attack increased its effectiveness by swamping the enemy's recovery services.

A feature of these formation raids was our involvement with other Liberator squadrons, not only of RAF 231 Group, but also with the gleaming aluminium Liberators of the USAAF, usually attacking from high above us. As to the identity of the other RAF Liberator units: in addition to 355 Squadron, 99 and 215 Squadrons regularly joined us in these attacks. So that, comparatively speaking, they became mass affairs, as sixty or seventy aircraft funnelled into the target area and individual squadrons found their allotted place in the attacking stream: each squadron sporting its distinctive tail markings for easy identification.

When pattern bombing worked, and usually it did, we felt enormous relief and, dare I say it, satisfaction. When it did not, and the pattern was partly or largely off the target, there was anguish, frustration, and – on the intercom at least – some barrack-room language. During their fall, bombs are affected by the wind, and if the air bomber in the leading aircraft had made an error in assessing the mean wind speed and/or direction, from a typical bombing altitude of eight thousand feet, the error on the ground could amount to hundreds of yards.

The policy of flying in formation on our daylight raids was not for offensive reasons alone. A single Liberator, in spite of its generous defensive armament, would have been a dubious match for a formation of Japanese fighters. But a compact formation of eight or twelve Liberators, with their combined firepower, was another matter. (From early 1945 we also enjoyed occasional top cover from Thunderbolt fighters; but such luxury was confined to short-range targets, such as those in the Rangoon area).

Thus it was that, although attacks on the Burma-Siam railway (invariably daylight affairs) were made by individual aircraft, we nevertheless flew out in formation and regrouped, where necessary, for the daylight portion of the return. Our targets were bridges and locomotives, which we bombed from three hundred feet or so, using delay fuses of 11 seconds to allow us to get clear before the bombs exploded. The major problem was persuading the bomb to stick on impact, and it took a while to become at all proficient in the art. The instinctive approach of putting the aircraft practically on top of the target at a few feet, before releasing the bomb, merely brought gasps of astonishment from the rear gunner at the sight of the bomb bouncing along in pursuit of him: initially achieving heights well above his turret, as we sped away, hugging the ground to evade the defences.

The approved method of attack from low level was to regard the target as the runway threshold in an airfield circuit pattern and to approach via a 'downwind' and a 'crosswind' leg; taking care to keep a safe distance from the aircraft ahead, so that its bomb exploded comfortably before one closed in. By arriving at the release point with the aircraft in a nose-down attitude, our bombs – stowed horizontally – were at least given the right bias as they left the aircraft: the theory being that during their fall they would become further nose down and so stick on impact. Sometimes it worked.

The Americans, rarely short of a solution in such matters, fitted spikes to the noses of their bombs, and it was soon apparent that their low-level bombing results were superior to our own. As with the mini-control columns fitted to the auto-pilots of the USAAF

Liberators, we were never supplied with this modification, but at least someone on our side benefited from it.

Our rather studied low-level bombing technique, in which we delivered our bombs singly – as opposed to dropping them in one or two sticks – did however invite some added risk. By the time we had queued up to deliver our third bomb, the defences were usually drawing a bead on one's aircraft. And although we encouraged our gunners to rake the defensive gun emplacements with machine-gun fire as we ran in to attack, we nevertheless sustained a fair amount of superficial damage during such operations, as well as losing an aircraft over the target on one occasion. In later attacks we tried dropping our bombs in a single stick flying straight-and-level at five hundred feet, which was fine when he hit the target, and maddening when we missed.

The few attacks we made at night were almost exclusively against the Makeson railway workshops on the outskirts of Bangkok: a target of strategic importance to the Japanese lines of supply. Being deep in enemy-held territory, it was both beyond the range of a daylight formation attack and altogether too risky a target for single aircraft penetration by day. In this instance, our tactics were to employ between eight and ten aircraft, taking off at two-minute intervals in the late afternoon, with the aim of achieving a stream attack concentrated within the space of twenty minutes or so. In time, 231 Group adopted a crude form of target marking for such raids, of which I add something later (see pages 135 and 136).

A Chance To Lead

When Lord Louis Mountbatten arrived in South East Asia in the newly created post of Supreme Allied Commander, he quickly made his presence felt by addressing the fighting units in person. He would stand on a piece of furniture – on hand for the purpose – and invite the troops to break ranks and gather round. His brass hat, charisma and fighting talk did the rest. It had worked for Monty in North Africa, and one sensed that Lord Louis recognised a good thing when saw it. Understandably he could not be everywhere, and he gave priority to visiting the ground troops, whose morale was most in need of stimulation.

However, those combat forces he could not reach in person were not to be denied the substance of these pep talks. Which was why, on 18th October, the AOC 231 Group (Air Commodore Mellersh) made the journey to Salbani to address 355 and 356 squadrons before a maximum strength raid on the port facilities at Moulmein.

The detailed briefing was allowed to proceed as usual, after which the AOC mounted the stage and said his piece. We were left in no doubt that the 'Supremo' personally regarded our mission as one of great importance: in infantry language, the word from the noble Lord was to get stuck in. Which, given the clinically professional nature of our normal briefings and the fact that no one needed to be told that it was important to hit the target if at all possible, gave the AOC's part in the proceedings a rather contrived, faintly theatrical, air. I'm sure I was not alone in feeling slightly embarrassed for him. Our objective was the destruction of supplies staging through the Moulmein trans-shipment sheds; important to the enemy in the forthcoming ground fighting. Important yes, but one which justified the role played by our unfortunate AOC? At what stage would they get round to issuing the rum, one wondered.

After take-off however, our immediate concern was how to link up into effective fighting formations. A solid layer of cloud extended from a few hundred feet above the airfield to three or four thousand feet. The brief for wing men, like myself, was to join up with one (any) of our appointed formation leaders en route: for which purpose they would attract our attention by firing Very lights.

Almost an hour after departure, and well out over the Bay of Bengal, we spotted two other Liberators. But since neither fired Very lights, I concluded that they too were wing men. This scenario had not been covered in our briefing, and I decided to offer myself as their leader; at least until one of the official box leaders showed up. Within a minute or two of firing our Very signals, both aircraft had joined us, and shortly afterwards another wing man tagged on to complete a box of four.

This new situation was at once to my liking. In the lead, I did not have the sweat of keeping station on someone else, and I had the satisfaction of deciding our route through the weather. More important, although the height of attack had been set out in broad terms at the pre-flight briefing, I would be the final arbiter in such matters as far as our small section was concerned.

Our squadron had recently received its first intake of airbombers, and for Joe Carberry, our airbomber, this was his first operational flight. As the lead airbomber in our small formation, he was about to be pitched in at the deep end. To begin with, assisted by Jock, and with inputs from Ted Holmes measuring the drift from his rear turret, he would have to compute the wind to set on his bomb sight; and on the bombing run itself he would play the key role. On the surface, Joe seemed a very self-effacing character and a bit uncertain of himself, and as we approached the target I could not help thinking that the next half an hour might be quite revealing.

Our recommended bombing height was in the region of six thousand feet. But as we let down to this height, I could see that Moulmein had a layer of cloud directly above it at about three and a half thousand feet. The question was, would the cloud be sufficiently broken for us to bomb accurately from above it? What was more, with the remainder of the Salbani wing and the USAAF Liberators also due to run in, I reckoned ours had to be a single run. I decided that to bomb from above cloud would be to risk failure, and I dropped down to three thousand feet. It would be a low run, and we could expect strong defensive gunfire. But the target would be clearly visible, and by coming in that low, we might even catch the defences off guard.

Some of the larger port installations were now visible: there were just under four minutes to run. I dropped down another five hundred feet to make certain we would be well clear of the cloud base. Just then Joe said his bomb sight was playing up and asked if I could hand over the lead to one of my wing men. What a moment to discover a snag! In the briefest of exchanges, I persuaded him to isolate the gyro mechanism and to do the best he could using the sight in its emergency mode. To have attempted a change in leadership at that crucial juncture could have resulted in chaos. Given the timing of Joe's announcement, it was a simple case of choosing the lesser of two evils.

With a minute to run, I glanced at the wing man on either side. They were in tight formation and rock steady, and my rear gunner reported the fourth aircraft in very close line astern. We had already opened the bomb doors and Joe was calling a series of minor corrections to my heading. It sounded promising. The trans-shipment sheds loomed up, and suddenly the sky around us broke into a rash of black smoke puffs, as the defences let fly. Meantime Joe was calling, 'Left; left; steady; steady; steady; bombs gone.' From the bomb-bay intercom Terry Malcolm confirmed, 'Bombs have gone, Skipper.' A short agonised wait and Ted Holmes announced with a jubilant shout that we had scored a direct hit: which, given Ted's, natural reticence, was like champagne. We had done it! Glory be! It was time to be off.

A quick check left and right to see that our formation was intact, and we were heading out over the sea, leaving the target area as swiftly as possible. Our wing men were still in tight defensive formation, and at this point, I allowed myself a quick glance over my shoulder. A thick pall of black smoke was already rising from the trans-shipment sheds. Our formation had been the first to attack, and if the others were as fortunate, Lord Louis should be more than satisfied. Just then, high above us, we saw two boxes of silver Liberators running in: the USAAF had joined the party.

The members of my makeshift formation elected to stay with me throughout the return to Salbani, where they went into starboard echelon and peeled off at intervals for an immaculate stream landing. It was a copybook performance on their part and, for me, it put the gloss on a thoroughly satisfying day.

Our debriefing, like the briefing before the raid, bore the Mountbatten stamp. Sitting in the background whilst we answered the Intelligence Officer's questions, was a British War Correspondent, who as soon as the official debrief was through, pounced on my crew for a story. This sort of thing was quite new, but having received an assurance from the Intelligence Officer that it had official blessing, we regurgitated the bones of what we'd been up to, and,

somewhat self-consciously, parted with details of our personal nick-
names, next of kin, home addresses and so forth. One way and
another, the Supremo was clearly determined to put his theatre on
the map.

As we were waiting for our debriefing, one of my wing men had
expressed mild surprise that we had gone in quite as low as we
had, but none of them seemed unduly perturbed about it. And no
one could argue with the fact that, unserviceable bomb sight or not,
Joe Carberry had done an excellent job. However, when Hugo Beall
walked purposefully towards me in the Mess that evening, I
wondered for a moment whether I was about to be ticked off for
grabbing rather too much initiative, or for going in too low, or
whatever it might be. I thought of the briefing of the previous
afternoon and the Supremo's message about the importance of the
target: well, at least I had something to fall back on. But to my
relief, Hugo merely wished to express his satisfaction with what we
had done.

The following month a photograph of our formation's bombing
pattern made the front page of the *Eastern Air Command News*, and
Hugo marked the occasion by promising me a section leader's position at an early date: a promise he duly kept. A fine thing, chance.

Unwinding

In our busier months we flew between seventy and eighty hours,
which, given our isolated location, left little time or opportunity for
socialising. Not that we didn't try: once in a blue moon a dozen or
so nursing sisters came to the Mess for drinks and a spot of dancing
to our wind-up gramophone; and off-camp, a party of us motored
twenty miles south to the Midnapore Railway Officers' Club on a
couple of occasions, where we cooled off in their swimming pool,
played the officers at billiards, made polite conversation with their
ladies, and drank their gin. But for ninety per cent of the time, we
relied on local resources and our own ingenuity to find a way of
unwinding.

A few people played bridge. Personally, I preferred poker – a game (as I discovered) at all times absorbing; sometimes exciting; and one which leaves the brain wholly untaxed. Our regular school – comprising Ed Hampton and Red Pierce from Canada, Mac McCormick from Australia, Paddy Carson from Ulster, Ron Haines from London, and myself – assembled most evenings and played through from after dinner until eleven o'clock. The stakes were nominal, but in spite of that there was keen rivalry to come out on top. To begin with I was very much the innocent of the school, but under the shrewd influence of Ed Hampton – a colourful character, who seemed to know every variety of stud poker – I soon had the rudiments. That did little, of course, to help one read the other man's hand. That element of mystery is what makes the game. It was a case of gauging from his attitude and facial expression whether or not he really did have the other king or whatever. By sitting down with the same five people on a regular basis, small give-away mannerisms, which had preceded an earlier bluff or strong hand, sometimes emerged. Or did they? Often one got it entirely wrong.

Apart from an occasional party, when we really went to town, our consumption of alcohol was extraordinarily moderate. First, because our personal ration of two quart bottles of beer per month could be consumed without the least difficulty in the course of an hour. Next, the spirits were practically all of Indian origin, and to an inexperienced palate certainly, they had a fierce, literally breathtaking character, which in itself put one off to a degree. Then there was the unwritten rule making alcohol taboo the night before operations, which also imposed some restraint. And finally, to some extent, flying in itself got rid of those tensions and frustrations which we might otherwise have doused with drink.

Regular station entertainment comprised little more than the nightly cinema show and a game of soccer in the early evening. But in spite of that, station morale remained remarkably high; in contrast with our earlier experience at Kolar. And since both forms of entertainment had also been available at Kolar, it was natural to question the difference. Personnel apart, one had to attribute this to the spirit engendered by Salbani's active involvement in operations: a role which made heavy demands, but which also welded us into a team and gave a lot of satisfaction. To be completely fair, our domestic accommodation at Salbani was somewhat less basic than that at Kolar, and certainly a weekend in Calcutta had the edge over Bangalore for letting rip: but I believe such considerations were mere icing on the cake.

Participative sport, except for one or two officers who played soccer, and, at a stretch, my neighbour's passion for tossing horse-shoes, was much neglected by our Mess members. But given the debilitating climate and the lack of facilities, perhaps we could be forgiven. A station swimming pool, on the other hand, would have been a godsend and much used, I'm sure; but although there was much talk of building one, it failed to materialise during my time at Salbani.

Moving off the camp for a long weekend, which we did about once a month, the choice was automatically Calcutta: a city teeming with people, and compared with Bombay, patently more squalid for people at the bottom of the ladder. But if one could stomach that, and was not put off by the ever-present begging hand and the constant appeals of 'Sahib, sahib', Calcutta could certainly provide relaxation and amusement. The Saturday Club, very English and a bit stuffy at first, was an oasis of calm and civilisation; while other, less staid establishments, were more than ready to add the spice.

Whenever we went to Calcutta as a crew, although by mutual consent we spent much of the time going our separate ways, we invariably set one evening aside to have drinks and dinner together. The Great Eastern Hotel in Dalhousie Square became the accepted venue for these occasions, at which we contrived to let our hair down as the evening went on. Chile usually gave us a rendering of his national anthem, entreating us to observe the dignity of the occasion, and insisting on the rest of us joining him in a toast to his native country. Red Rigby's contribution consisted of several verses of *Albert and the Lion*. And so it went on, until everyone had done his turn. Not desperately exciting as cabarets go, but enough to keep us amused at the time. I can't recall my own contribution, although it could have been an attempt to sing *We do come up from Zummerzet*. As a rule my real party piece began as we *left* the hotel.

Ordering a horse-drawn gharry, it was my habit to mount the box, and cajoling the reins from the driver, to drive it down Chowringee – Calcutta's busy main street – and on to our hostels. Normally this went like clockwork. But one evening, attempting to turn round in a narrow street with a marked crown to it, my shortcomings as a gharry driver were rudely exposed. Having got the horse well into the turn, the gharry wallah suddenly became quite agitated, pleading to be given control. Failing to grasp the problem, I'm afraid I brushed aside his protestations and proceeded with the turn. The next thing I knew the gharry was on its side; the horse was threshing its legs in the air; and the driver was castigating me for being the idiot that I undoubtedly was.

A crowd quickly gathered, and we began to take some abuse; verbal at that point, but things looked as thought they might easily turn nasty. (As the gharry was about to keel over, my crew had smartly abandoned ship and were now awaiting my next move.) I was not proud of the mess I had caused, but this was no time for indecision or to linger over apologies. I thrust what I felt would cover both the fare and any damage done into the hand of the tormented driver, and we made off at the double to some loitering rickshaws. With the noise of the jabbering crowd still uncomfortably close, we cut short the normal haggle over the fare and naming our destinations, leapt aboard; ordering them to *Jaldi! Jaldi!* (Quickly! Quickly!). We were half a block away before my pulse was back to normal.

Longer breaks from duty depended essentially on the squadron's commitments. But most aircrew managed to get away for a fortnight's 'privilege' leave during their operational tour, even if they had to accept pot luck on timing: the situation for our ground staff being similar. Mostly we headed for the hills, to Kashmir or Darjeeling, where cool air and green surroundings could be guaranteed, and some feminine company was more than a possibility. If there was a next time and one wanted a change of scene, there were the seaside resorts of Puri and Pondicherry on the Bay of Bengal where, during the winter months at least, a cool on-shore breeze and a warm sea offered a relaxing combination.

The transformation worked on people during those longer periods of leave was remarkable. Short tempers, anxieties and tensions over minor issues, even the ravages of septic prickly heat, just ebbed away; and colleagues who one felt had not smiled in weeks were suddenly capable of a broad grin. The result was restored morale and increased efficiency. Higher authority's policy of getting the crews off the plains and into the hills now and then was more than an act of kindness.

Fighters! Fighters!

On 22nd October, we flew as the number two in a box of four aircraft to renew the attack on the port of Moulmein. The squadron put up between twelve and sixteen aircraft that day, and our box was part of a much larger formation. To eliminate unnecessary fatigue, we were briefed to follow the standard practice of flying in loose formation whilst in friendly airspace – except when penetrating cloud – but over enemy territory to tighten up for mutual protection.

The weather was bright and clear, except for some well broken strato-cumulus; to the extent that, en route to the target, I remember feeling quite exposed without the cloud cover which had masked our approach on earlier raids. As we neared the target, we again saw American Liberator formations high above us, also heading for Moulmein, and I wondered what super bombsight they were using, that they could be so confident of hitting the target from that altitude. We found it difficult enough, from thousands of feet lower. I also found myself envying them their superb station keeping, which I attributed, rightly or wrongly, to their special fingertip controls which I mentioned earlier.

Over the target we encountered heavy anti-aircraft fire: several of our machines, included my own, receiving superficial damage. Our station keeping, however, remained good, and it was annoying to hear Joe Carberry report that the whole of our formation's bomb pattern had overshot the target. As usual, we had followed the practice of taking our point of bomb release, as well as our line of attack, from the overall leader. On subsequent pattern bombing raids, as a result of this débâcle, each individual air bomber calculated his own point of release, so that we never again risked compounding an 'error' of that sort on the leader's part – whether it was due to human miscalculation or an equipment malfunction. Some lessons were learnt the hard way.

By the time the enemy fighters struck, we were already well out over the Gulf of Martaban on our return journey, and our formation had begun to loosen up. Suddenly there was an excited cry (from Ted Holmes) of 'Fighters! Fighters!' Nothing resembling numbers, position or range. 'Where? How many?' I snapped back. Meanwhile, with the adrenaline flowing, I had slammed open the throttles to get back into close formation. 'One fighter, nine o'clock, level, three hundred yards' came the reply. Glancing left, there it was – a single, radial-engined machine, jungle green in colour, not closing in to attack, but performing aerobatics of all things.

Just then, with no more than a few wing spans now separating our aircraft, two Japanese fighters ripped through our gaggle from above, spraying us with machine-gun fire as they did so: the joker on our left had been a decoy. By now the adrenaline was really buzzing. Any moment, surely, another attack would come. But could I get back into my position? – no, someone else had grabbed it. Question – should I slot in just anywhere or might this lead to misunderstanding and a possible collision? I felt it might, so I closed alongside my lodger, inviting him to give way. He did, but reluctantly, and later we had words.

No further attacks came, but we had learnt our lesson – for the remainder of our passage through enemy air space, hardly a wing span separated us. Fortunately, none of our crews were injured by the attack, and damage to those aircraft hit was minimal. The Americans had been less fortunate however: shortly after the attack, we saw two pillars of dense black smoke rising from the waters of the Gulf.

For their cheek, bravado and professionalism, it was impossible not to admire our attackers: for the way the decoy had positioned himself down-sun of us, nicely out of range of our guns, and drawn our attention with his aerobatics, while the rest of his formation manoeuvred to attack us from above, and out of the sun. And the attack itself, in which they had not pussyfooted around on the edges of our formation, or concentrated their attention on a straggler, but bored straight through the middle of our loose gaggle. Had they been equipped with cannon, they could hardly have failed to nail one or two of us. Or maybe they *were* equipped with cannon and had used up that part of their ammunition on those unfortunate Americans.

Being attacked by fighters twenty minutes or so *after* delivering our bombs was hardly something we expected, but the underlying reason for loosening up our formation too soon was the general absence of enemy fighters over Burma at that time and the lack of vigilance this had begun to breed in us. To preserve their aircraft from Allied attack, the Japanese normally kept them well south of our daylight radius of action; bringing them forward only on an ad hoc basis. The very fact that we were attacked well out from the target on our way home (classically, the role of the fighter is to destroy the bomber before it reaches its target) was almost certainly a result of their operating from a rearward base.

This aerial encounter reflected little credit on us, revealing as it did weaknesses in our tactics and procedures; and we could certainly count ourselves fortunate in receiving a salutary lesson at virtually no cost. Our deficiencies went beyond bad defensive formation keeping, for which I accept my full share of blame. Speaking for my crew at least, there was Ted's lapse of composure when reporting the presence of fighters so that for a few precious moments the rest of us had no idea of numbers and position etcetera. Finally, there was the confusion and delay caused by someone grabbing my position and insisting for some time in holding on to it.

It was routine in 356 Squadron to hold a post-operation critique the day following a raid, by which time people had slept on their grievances and tempers had had time to cool. As a result, this normally amounted to the formation leader and his airbomber giving a dispassionate resumé of the attack phase and the bombing results, together with any advice they might have to offer on improving our techniques on future raids. The discussion period normally fizzled out pretty quickly; that is, if it got going at all. But following this raid, it found no shortage of speakers or absence of fervour as the issues were hammered out.

Leaders and Personalities

With a dozen or so operations to its credit, 356 Squadron had become more than a number plate on the Air Order of Battle – our leaders had been tested and personalities had emerged. Hugo Beall – who as our Commanding Officer led the majority of the early raids – had come through as a competent and determined leader, whose devotion to the job was absolute. His airborne professionalism being backed to the hilt by his insistence on both thorough pre-raid preparation and free-for-all post-flight critiques.

Given that he was a person of such singular devotion, it will, perhaps, be no surprise that initially Hugo came across as a very serious, even sombre character. When this short, trim, unsmiling man with the piercing blue eyes approached, we often found ourselves examining our consciences. His stiff, purposeful walk in itself radiated a certain air of tension, reminiscent of a tightly wound spring.

Those close to Hugo – like his Adjutant and his crew – however, were aware of a warmer man. And the longer we served under him, the more we realised that with him loyalty was very much a two-way street. He expected and he took, but he also gave. Hugo's ability to instil loyalty, combined with the squadron's increasing success and his obvious sense of fairness, ensured that during his stewardship morale steadily rose; so that when he was awarded the Distinguished Service Order on the completion of his tour, no one doubted it was thoroughly deserved.

Our senior Flight Commander (OC 'B' Flight), and the person who occasionally stood in for Hugo in leading our daylight attacks, was Squadron Leader 'War' Harris, a very correct man with a clipped military-style moustache and a slightly high-pitched voice. In fact his gentlemanly style and bearing, together with his rather reticent manner, was more in the mould of a country squire than that of the wartime concept of an RAF officer – outgoing, rather noisy and a bit of a lad with the women. I suspect War began his service life in the army, transferring to the RAF after service in an Army co-operation squadron. If so, it might go some way to accounting for his pukka ways.

But there was nothing reticent about War's behaviour in the air. He quickly proved himself an able leader, and was much respected for his ability to lead large formations in a way that gave every pilot in the formation a proper chance of holding his position, whatever manoeuvre the situation called for. When such leadership is repeated over a series of operations, wing men flying on the extremities of a squadron formation know that it is not just a matter of luck on the leader's part. They come to realise, better than anyone, that he was able to read the situation – the weather especially – and by sensible anticipation to avoid the last minute, large alterations of course, altitude and/or speed, which sometimes distinguished the less able leader. With the Liberator's limited speed range, and the general need to conserve fuel, such considerations were of real importance in maintaining the cohesion of a large formation.

My crew was assigned to 'A' Flight, where in our first five months, we averaged a change of Commander every other month, which hardly made for continuity. Two good men – Don Ritchie (Australian) and Johnnie West (Canadian) – came and went: they were caretakers, it seemed. From there on, however, after Eric Brighouse had been given the acting rank of Squadron Leader and was in the chair, we did enjoy a five-month period of continuity. Eric, or 'Brig' as he was known, was a Yorkshireman in his early thirties, articulate and shrewd. He had begun his aircrew service as a wireless operator/air gunner, which meant that compared with War his experience as a pilot was relatively limited. As a result, he was not in quite the same league when it came to the niceties of handling large formations, but there was no doubting his leadership in terms of resolution.

There was no denying either that Brig was a character, and great fun socially. At a party, particularly if there were women present, he was an enormous asset in getting things moving. His portly figure,

mischievous wit and rollicking peals of laughter seemed to act as a
magnet to the more matronly among them. But there was also a less
flippant and artistic side to his nature. Before the war, he had trained
as a draughtsman, and his love of line was reflected in the copperplate
hand with which he studiously entered our names in the flight
authorisation book before each operation. Given that this sometimes
involved sixty-odd names, it almost amounted to a labour of love.
(When I last met him in the 1960s, he was designing cabinets for
one of the big London furnishing stores – and very handsome
designs they were too.)

The Gunnery Leader, Flight Lieutenant 'Robby' Roberts, was
one of those larger-than-life individuals. In temperament and
demeanour he was almost the antithesis of War Harris. A fine physical specimen, standing a shade over six feet, and with a permanent
roguish smile, he was a wild, rumbustious character – the complete
extrovert and a man of restless energy, but an energy which was
frequently misdirected and which occasionally got him into scrapes
with authority – as when a hail of shotgun pellets rained down on
the roof of Hugo Beall's bungalow one afternoon when Robby opened
fire at some birds. Hugo was having a siesta at the time and was not
a bit pleased.

Neither was Robby in the good books of our local 'Works and
Bricks' (MPBW) representative. In this instance, having tired of
making do with the six inches of water to be had in his standard-
issue canvas bath, Robby built himself his own full size bath with
bricks and cement, locating it on the back verandah of his bungalow:
no planning permission or request for materials; he just gathered
the latter from around the camp and went ahead. A very fine bath
it was too, getting on for five feet in length and complete with plug
hole for easy drainage. But apart from the indignant reaction from
Works and Bricks, I'm not sure that Robby's bearer was too pleased
with the result either, given that he had the job of filling it every
day.

Robby's routine job was to keep our gunners up to scratch in
their ground training and in the regular inspection of their turrets.
But prior to each raid, he also spent a good deal of time ferrying

belts of ammunition from the armoury to the aircraft dispersals, which apparently entitled him to the permanent use of a three-ton lorry and another outlet for Robby's wild streak was regularly to drive this vehicle between the Mess and the airfield at such breakneck speed that after a while no one would accept his offer of a lift, given a choice.

As to Hugo's reaction to Robby's antics, I think he was shrewd enough to realise that Robby could not be tamed and that the best he, Hugo, could hope to achieve was to channel Robby's surplus energy into a useful supernumary task. That, I believe, is how Robby came to take on the job of messing member, which apart from resulting in a distinct improvement in the standard of the Mess food, served to keep him out of harm's way for at least part of the time. But being anywhere near Robby remained a bit like driving around with a time bomb in the boot of your car: you just knew that the peace of the moment was not going to last.

When War Harris completed his operational tour, Squadron Leader 'Ned' Sparks assumed command of B Flight, and, in due course, when Hugo's tour expired, took command of the squadron. Like Hugo, Ned was a Canadian, but unlike Hugo, who was a member of the RAF, Ned was of the RCAF. With replacement crews shortly to become predominantly RCAF, Ned's appointment made good sense. But politics apart, he was also the obvious choice in having the right combination of seniority, experience and time to serve.

Ned was a wiry, mature man in his early thirties, with a small moustache, neatly cropped hair and with something of the head-master about him. As an aviator, he was the complete professional: concerning himself with every detail. As to Ned's leadership in the air, I left shortly after he took command and cannot speak from personal experience, but I have not the slightest doubt that it was of top quality, since he too was awarded the Distinguished Service Order.

In July 1945, 356 Squadron was redeployed to the Cocos Islands to take it within better striking range of the retreating Japanese.

Sadly, within weeks of making the move, Ned went down with poliomyelitis, from which he died within days.

Squadron Leader Les Evans, the man who assumed temporary command on Ned's death, was of a more relaxed temperament, although entirely sharing Ned's professionalism. I left the squadron well before the move to the Cocos, but I would be surprised if in taking command Les had shed any of his sunny disposition and his gift for subtle persuasion: characteristics for which I remember him when we served under Eric Brighouse. In terms of experience too Les had a considerable edge over people like Stringer-Jones and myself, both of us straight-through products of the training machine. When Les arrived at Salbani, he had already distinguished himself as a flying instructor on Masters and Oxfords: work for which he had been awarded the Air Force Cross.

Les assumed temporary command on 10th August 1945, and a few days later 356 Squadron flew its last mission before VJ day: the task – a sea search for an aircraft missing from 321 Squadron. Wing Commander Walker who was posted in as Ned's replacement did not remain on the Cocos for long before joining Lord Mountbatten's staff, leaving Les in charge to wind up the squadron. The value authority put on Les Evans's work with 356 was also to award him the Distinguished Service Order: as with the awards to Hugo and Ned, well received by the squadron members, I'm sure.

My object has not been to make this short chapter a roll call of all the squadron's characters or of those who served 356 with distinction – officially recognised or not – and I hope the fact that it stops short of mentioning a number of people who deserve to be named will not give offence. But there are just two more people I feel I must mention, however briefly.

Flight Lieutenant Ted King, the Engineering Officer, and Pilot Officer 'Ginger' Caldicot, the Armaments Officer, both worked minor miracles on occasions in solving problems against the clock, and in maintaining the morale and efficiency of their men in spite of monsoon downpours, oppressive heat, and occasional dust storms. Without the sustained effort of Ted and Ginger and their men, the work of we aircrews would not have amounted to a row of beans.

Finding the Target

On any but the short-range sorties of six hours or so, we were obliged to restrict our bomb load in order to carry enough fuel. The more distant the target, the greater that restriction, and the more imperative it was to bomb with precision.

But first, we had to find the target. Easy enough if it was a sea port, an airfield or a large railway yard. Not so easy if it was a camouflaged stores dump in fairly open country, or a particular bridge in an area with as many rivers as you have lines on the palm of your hand, and umpteen bridges to match. If it was a night attack, moon or no moon, the problems of identifying the target were normally compounded by the reduced visibility.

Operating over a radius of action in the region of a thousand miles, with much of that over sea or featureless jungle, meant that quite large errors could build up in our dead-reckoning navigation en route to the target. On the other hand thanks to a powerful homing beacon at the USAAF B29 base at Kharagpur, just south of Salbani, the final stages of the return flight held no such problems. During my service with 356 Squadron, we had no long-range navigation aid, although something was on its way. (Receivers for LORAN, a ground-based long-range aid, were installed in all but two of the squadron's aircraft during May 1945, and were ready for operational use by the end of that month.)

At night we used astro-navigation – Jock Hunter asking me to hold a steady course, as he prepared to wrestle with his sextant. Crossing the sea, night or day, we dropped markers from which to measure our drift and calculate changes to our heading. Sometimes the weather prevented these activities, and we were back to the triangle of velocities and dead-reckoning.

Entering the general target area, our method was to head initially for a predetermined and distinctive ground feature a short distance from the target itself – say ten to fifteen miles away. This technique of approaching the target from a precise position (the Initial Point or IP) avoided large alterations of course on the bombing run itself, and reduced to a minimum the risk of a mistake in identifying the target. Before take-off, our airbombers made a minute study of the map features on the run-in, so they knew, within the limitations of the map, what to look for as each minute ticked by.

But not infrequently, the maps themselves were inaccurate in matters of detail, and this in itself could cause delay in reaching the target. This was especially so at night, when darkness limited our field of vision – sometimes radically – and denied to us the use of other features in the vicinity from which to confirm our position. An extreme example of map inaccuracy cropped up during a sortie we made over French Indo-China – nowadays Vietnam – where we found a mountain towering well above any high ground shown on the map. Another trap for the unwary, was to accept without question the map maker's portrayal of the size and shape of rivers and estuaries: features which could vary significantly, depending on the season and, in the case of the estuaries, on the state of the tide as well.

In conditions of poor visibility, there are of course, accepted map-reading techniques to minimise its effects, and we made full use of these; in daylight for instance, having one's back to the sun when looking for ground features and, conversely, at night, facing the moon when doing so. Also, at night, the pick-up range of ground features can be increased significantly by reducing cockpit lighting and allowing sufficient time for one's eyes to adapt to the outside darkness before making any serious attempt at map reading.

But, as we discovered, success in finding the target was often a matter of employing every aspect of technique, and of not relying too heavily on any single method.

A Long Night

In the late afternoon of 2nd November 355 and 356 Squadrons each despatched five aircraft on a night attack against the Makeson railway workshops on the outskirts of Bangkok, the Siamese capital. Take-offs were at two-minute intervals; and inside twenty minutes the whole force was on its way with the exception of my own aircraft. One of our engines was badly below power; and as the last aircraft lifted off, my groundcrew were still feverishly engaged in a plug change.

A few minutes later Hugo Beall drew up in his jeep: his instructions brief and to the point. If I was unable to get into the air within the next twenty minutes, I was to cancel my mission. He did not want me any further behind the main concentration. I saluted, and our groundcrew did the rest. Twenty minutes later Jack Clarke was raising the undercarriage as we built up speed after take-off.

It was a beautiful evening as we crossed out over the Sunderbans south of Calcutta: the waters of the Bay of Bengal flat calm and shimmering in the evening sun. The inter-tropical weather front was well south and, with luck, we would be clear of cloud most of the way to the target. Almost too clear perhaps. I glanced at my watch, seeking reassurance from Jock Hunter that it would be properly dark before we reached enemy airspace. Jock said it would.

We made a good landfall west of Bassein and, having crossed the Gulf of Martaban, we made another one on the Tenasserim coastline. Our original brief was to pinpoint our position on the Gulf of Siam coastline, and to run in to the target from there. But as I saw it, if our colleagues ahead of us had done their job properly, we should be able to make a direct approach onto the fires started by their incendiaries. So it proved.

We first saw the fires from thirty to forty miles away. Clearly the target had already received a fair pasting: later we learnt that fifteen aircraft of 159 Squadron had joined in the attack. Nearing the target, I disposed the crew to their stations for the bombing run; reminding Joe Carberry of the need to identify the target as such, and not merely to aim at the fires.

We were still some five minutes from the bomb-release point, when a number of bluish search-lights stabbed the darkness, which judging from the immediate accuracy of their aim, appeared to be radar guided. Almost immediately we were subjected to a lively barrage of anti-aircraft fire, which persisted with some intensity until we had released our bombs and were some minutes outbound from the target.

On the bombing run itself, the noise of the AA shells had sounded like a giant whip lashing the fuselage. There was also a good deal of turbulence and, at the time, I also attributed this to the gunfire: on cooler reflection, it was almost certainly due to the heat from the fires below. Nevertheless, it was all rather distracting. The more so, no doubt, because the defences had only one target to deal with.

Clear of the Bangkok defences, I asked Ted Holmes and Red Rigby if our bombs had hit the target: Ted, because the view from the rear turret was the equivalent of a seat in the grand circle, and Red because, amidships, he had a splendid view out of either beam hatch. At first, both seemed vague about our results. But Red finally said he was sure the fires had been stoked up by our attack. And from his bomb-aimer's station, Joe confirmed that his release indicator had registered the proper release of our bombs.

I was a bit disappointed not to hear more conclusive evidence of our success but decided to wait until Bangkok was well behind us, before pursuing a more explicit report. The accuracy of the search-lights and the gunfire had surprised me, and I thought it more than possible that such sophistication might be accompanied by the presence of night fighters. I impressed this view on the gun positions, and Jack and I turned our cockpit lighting right down, so as to black out the aircraft as far as possible. This in spite of the fact that, viewed from below and behind, our white-hot engine turbine discs would have made a wonderful homing beacon.

Well out from the target – almost an hour on our homeward journey – I felt I could indulge in a more detailed chat about our bombing results. At this stage Ted was still in his turret and Red was caressing his .5–inch machine-gun at the beam position. By now I had begun to entertain the nagging thought that we might have dropped our bombs without triggering their fuses. Otherwise surely, Ted or Red, if not both, would have seen the target erupt more positively. Ted maintained his reticence, and on this occasion, even Red could not find the words to reassure me.

I could wait no longer to dispel my doubts, and instructed Terry Malcolm (my wireless operator) to enter the bomb bay and physically to check that the fusing links had been retained when our bombs were released. If this was the case, then our bombs had indeed been

properly armed, and my doubts would be removed. The intercom
burst into life. It was Terry giving me a message even I had not
foreseen. Our bombs were still on board!

Pushing aside protests about lighting up the aircraft over enemy
territory, I had Terry switch on the bomb-bay lights, and turning in
my seat, I looked aft through the trapdoor into the bomb bay. Row
upon row of undelivered ordnance stared back at me. I was stung
into action.

First, I asked Red if we had enough fuel to return to the target
and still reach Chittagong, the nearest suitable base in friendly
territory. He thought we had. Next, whilst Red was coming forward
to fine-tune that reply, I turned the aircraft about to head roughly
towards Bangkok and asked Jock to work out a proper heading.
Before this turnabout we had almost reached Heinze Bay on the
Tenasserim coast, so Jock had a good pinpoint from which to correct
my heading.

At this point Joe came up with the idea of bombing a target on
the way home. I'm afraid my reply was more than a bit rough.
Provided we had the fuel, I was in no mood to compromise. Red
had now joined me on the flight deck, and checking his arithmetic,
I agreed with his assessment that we could make Chittagong with a
safe margin, enemy action permitting.

Fifty minutes later, the enemy defences gave us the mixture as
before: just as active, but mercifully still lacking that final degree of
accuracy. By contrast, our own methods represented an order
of improvement, as we contrived to remove every possibility of
another abortive attack: Terry was now in the bomb bay, squatting
on the catwalk and plugged into the intercom; immediately following
the bomb release, the bomb-jettison handle was operated; Terry
physically watched the bombs leave the aircraft, and confirmed the
fact on the intercom; and for good measure, on receiving Terry's
call, I reefed the aircraft into a turn and personally observed the
results. On this occasion our results really were good, although Jock's
log already bore a somewhat similar remark in an entry recorded
almost two hours earlier.

The remainder of the flight consisted of keeping the gunners alert whilst we were in enemy-controlled airspace; fussing over the engine settings and the airspeed to extend our range as far as possible; occasionally rechecking our fuel calculations; and, not least important, working with Jock to ensure the shortest feasible route to Chittagong.

At touchdown, we had been airborne for thirteen and a half hours: dawn breaking shortly afterwards as the aircraft was being refuelled. Looking at my crew, they were all hollow-eyed in varying degrees, and I'm sure I was no exception. However, over a bacon sandwich and a mug of tea, I managed to get a bit of banter going before we climbed back on board to complete our return flight to Salbani. This leg took another hour and fifty minutes, during which the general silence on the intercom betrayed, no doubt, not only fatigue, but also some concern about the sort of reception we might be in for.

Not surprisingly, our colleagues were now fast asleep, which spared us their ribbing until that evening. Meantime, the intelligence staff who debriefed us, and our flight commander, expressed relief at our safe return. News of our landing at Chittagong had still to filter through the ground channels, so until I called the airfield as we approached, we had been regarded as overdue.

As I reached my bungalow, Wallaya, who had been squatting on his haunches on the front verandah, got to his feet and grinning broadly exclaimed, 'Sahib late coming,' to which I replied, 'Yes, sahib late coming, but at least sahib *has* come.' We both laughed, Wallaya exposing a set of teeth bearing the red traces of betel nut. I remember thinking that this probably accounted for his unusually forward manner and his slightly tipsy appearance. However, since he'd had over three hours to kill since the other sahibs had returned, and no word about me meantime, it was hardly surprising that he should have relieved his boredom by indulging in his favourite narcotic. Although I got some sleep that day, I did not surface for long that evening before getting my head down again.

The next morning the Adjutant summoned me to parade with my crew in Hugo's office. Well perhaps Hugo felt he had no alternative to giving us a ticking off. But since the bomb-release apparatus had given no hint that it had malfunctioned on the first run, there was a strong element of bad luck about our abortive attack as I saw it. And having made a mess, we had at least cleaned up afterwards.

Escorted by the Adjutant, I led my crew into Hugo's office, saluted, and waited for it. Hugo, unsmiling, looked at me intently with those piercing blue eyes of his, and thrust a piece of paper into my hand. I was still pretty drained, and as I took it, my hand trembled slightly: from fatigue rather than apprehension. 'Here, Lucky,' he said, 'read this aloud – it's a signal from the AOC.' I was barely into the signal, before I got the point. We were being congratulated, not reproved: the whole squadron for a very successful raid, and my crew in particular for our determination to complete our mission successfully.

Hugo waited until I had finished, then getting to his feet, pumped my hand and, grinning broadly, added his own words of approval. I felt a bit embarrassed, but it was a good deal better than being ticked off.

Pause for Training

In December 1944, the RAF 231 Group Liberator force, with the exception of 159 Squadron, embarked on an intensive training programme; the objective in terms of individual aircrews being to reduce their bombing error to sixty yards. Given the relative pinpoint nature of our targets compared with, say, the bombing of area targets in German cities, only an accuracy of that order was considered acceptable. (In deference to our airbombers, they had arrived on the squadrons direct from the UK only three months earlier, at which stage none of them had received operational training or bombed from an operational aircraft, and only a small number of them had received previous instruction on the bombsight – the Mk XIV – fitted to the Liberator.) But there was more to the training programme than bringing our airbombers up to scratch.

With 215 and 99 Squadrons now part of the Liberator Air Order of Battle, there was also the need – with large formations of aircraft attacking the same target from widely separated airfields – to practise co-ordinated attacks, whereby the squadrons preserved the concerntration of the attack without getting in one another's way. Another important objective – if we were to play a useful role in support of the 14th Army's impending push – was the need to demonstrate our ability to bomb the enemy's battlefield positions without putting our own troops at risk.

In addition to the need to improve the bomb-aiming accuracy of crews generally, in the context of pattern bombing, the Group air staff also recognised the need to hone up the skills of the lead aircraft crews in particular. Consequently, the commanding officers of both 355 and 356 Squadrons, together with their navigators and airbombers, attended an ad hoc course of instruction prior to the Wing and Group phases of the training programme.

Three months earlier, the Group air staff had pursued the possibility of fitting a Norden bombsight in the lead aircraft of each squadron, in an attempt to ensure the greatest possible accuracy in our pattern-bombing attacks: this device being a highly classified piece of American equipment. The Group operational record book for January 1945 shows that three such sights were in fact delivered, but that trials were delayed due to the lack of certain parts: a logistics 'oversight' induced for political reasons, maybe? During my service with 356 Squadron, ending in late May 1945, I was not aware of these sights being deployed to any of the Liberator squadrons for operational use, but it is gratifying to see that our air staff had at least made the effort to acquire them. (Early in the war an American general was credited with saying that using the Norden sight a bomb could be placed in a pickle barrel from 10,000 feet. We at Salbani would have happily settled for being able to hit a three-ton lorry from half that altitude.)

The practice bombing programme began with sorties by individual crews, followed by exercises involving flight and squadron formations, and culminating in mid-month with a 231 Group formation comprising some fifty aircraft. In the latter – all aircraft operating from their home bases in Bengal – 184 Wing (356 and 355 squadrons) joined 175 Wing (99 and 215 squadrons) over Nagpur, and made simulated attacks in squadrons line-astern on Poona and Bombay, before proceeding to drop practice bombs on the range at

Bhopal: what a splendid sight that huge Balbo made, and damned hard work it was too. That day we flew for nine hours in formation before splitting up as night fell, and returning to our respective bases as single aircraft: even an exercise of those proportions being easily swallowed by India's vast airspace. Aside from how the enemy might react, the general consensus was that this Group exercise had simulated most of the demands we were likely to face in our future formation attacks.

A few days later, on 21st December, we put this training to good purpose, when twelve aircraft of 356 Squadron pattern-bombed a Japanese stores areas at Taungup in the Arakan, during which the target was blanketed by bomb bursts. An excellent result, which left us in great heart; entirely justifying the sweat of the training period. I know that 355 Squadron also took part in the attack on Taungup that day, and I suspect that 99 and 215 Squadrons were involved as well.

Despite the pause for training, the Group nevertheless continued to deal with certain categories of operational target throughout December, although in most instances at a much reduced rate of effort. Number 159 Squadron did not figure in the training programme because it generally undertook the more specialised duties, such as long-range mining and other night operations, neither of which were relevant to the training programme, which was exclusively devoted to daylight bombing.

December also saw some restructuring of the chain of command above the 231 Group level as the Allied forces in the area prepared to assume a more offensive role – Headquarters RAF Bengal/Burma being formed at Calcutta under the command of Air Vice-Marshal the Earl of Bandon; known to his friends and superiors as 'Paddy' and generally referred to (with some affection) by his subordinates as 'the abandoned Earl'.

A Working Christmas

Given the preoccupation with the practice bombing during the first three weeks of December, my operational hours for the month looked like being confined to my part in the pattern-bombing raid on Taungup on the 21st. So the opportunity to volunteer for a long-distance raid on Christmas Eve was a chance to more than double that total overnight. A brief consultation with my crew, and I seized it.

The target lay thirteen hundred miles to the east, just inland from the Gulf of Tongking, and comprised the railway sidings at Phulang Thuong, thirty-five miles out of Hanoi on the railway which runs north-east from the city. Penetrating that far into enemy territory dictated an attack by individual aircraft operating singly at night. To carry enough fuel – just short of three thousand gallons – I was obliged to leave behind my ball and mid-upper gunners, together with my airbomber and second WOP/AG. Even at that, our bomb load was reduced to a mere two thousand pounds.

We were to take off in the late afternoon, and as I rested on my *charpoy* (bed) after lunch, I picked up Palgrave's *Golden Treasury*, just to pass the time. As I recall Jack Clarke and I were the only persons in the officers' lines taking part in that night's raid, so that the usual feeling of togetherness was entirely lacking, and what with the nostalgic overtones of Christmas, and the sweetly seductive nature of the poetry, I suddenly felt my enthusiasm for the night's operation begin to ebb. The feeling was new; and vaguely alarming. I closed the book and sought company; not to open my heart to, but to put myself in a more extrovert frame of mind. A quarter of an hour of idle chat and I had pushed the feeling to one side, but Palgrave's remained unopened for some months afterwards.

With Salbani despatching a mere handful of aircraft that night, the intelligence hall lacked the usual buzz of the crowd, and with what seemed a straightforward task ahead of us, our briefing was through in short order. The met forecast was good; there were some well defined features in the general target area; and at the time of our attack, the moon would be there to help us pinpoint our target and to deliver our bombs with accuracy. The main challenge looked like being the sheer length of the sortie (thirteen hours or so), and as the aircraft captain, I was given a supply of caffeine tablets to issue at my discretion to help maintain crew efficiency: this being normal procedure for sorties of twelve hours or more – not that we used them much.

At the aircraft dispersal, I had completely shed my earlier mood of introspection and was looking forward to the flight. At this stage of events, one of our most experienced aircraft captains, before climbing into the cockpit, would invariably wander quietly into the surrounding scrub to be sick; such was his state of tension, poor chap. In my case, which was probably typical, the added tension merely served to sharpen my faculties.

Shortly we were heading east over Chittagong towards the Chin Hills and enemy airspace. As darkness fell, Jack and I reduced the cockpit lighting and sat back to await our crossing of the Chindwin; the first of the great rivers which straddled our track. If it was visible, the Chindwin would give us a useful check on our ground speed. And later, as we crossed the Irrawaddy and the Salween, we should, with luck, be able to fine-tune our ETA on target. Apart from wanting to maintain an accurate track, a key requirement was to calculate the time when we would be clear of high ground and safe to descend towards our target.

Beyond the Irrawaddy – which we crossed just to the north of Mandalay – our outbound flight was almost entirely over mountainous country, where spot heights of 8,500 feet were not uncommon. This put us at a cruising altitude of about ten thousand feet and meant we had to use oxygen from that point onwards, until we descended into the target area. It is one thing to get by at ten thousand feet without oxygen, and an entirely different matter to remain at that height for long periods and to continue to work efficiently.

Crossing the Salween, the cloud beneath us was sufficiently broken to give us a good check on our ground speed. But from there on, it gradually thickened. So that well before we had reached the time to begin our descent, it was a solid sheet. No one with a healthy interest in living likes the idea of descending through solid cloud unless he can point to his position on the map. And the fact that Jock could not account for the presence of a nearby mountain, which towered majestically above the cloud tops, added to my caution. My proposed solution was deliberately to overfly the target area, and let down over the sea. Once through the cloud, we could then double back, pick up the coastline – which should provide an excellent pinpoint – and map read our way to target. If this left us short of fuel for the return to Salbani, we could refuel at Chittagong. Jock and Red, whose specialist areas were involved, endorsed the idea.

We flew on for a while before making an uneventful descent. Over the sea the cloud was partly broken, and it remained that way as we made landfall at one of the mouths of the Red River. Locating the target from there was simple. We followed the river until we had the city of Hanoi in sight; then, skirting the outer fringes, we picked up the railway leading to our target. A primitive method of navigation perhaps, but effective. And one which I was not too proud to use, given the less than distinctive nature of the target itself.

On our bombing run, it was quickly evident that the prescribed direction of attack had been badly thought out. We were running in with the moon behind us, which quite unnecessarily reduced our forward visibility. And since the target demanded pinpoint accuracy, Jock – who was dropping the bombs in the absence of Joe Carberry – quite properly abandoned the run. When this happened a second time, I decided to ignore our briefing instructions and to make a run in the opposite direction. This meant keeping a sharp lookout for other aircraft, which we might meet head-on, but with the small number of aircraft involved, and the fact that we were rather late on target, I considered the risk acceptable.

As we circled to come in from the other direction, I caught sight of the shadowy outline of another aircraft following us, just out of gun range. As far as I could tell it was not another Liberator, and its stalking tactics were unsettling. I instructed Ted to keep it in his sights, and to deal with it if it attacked. A few minutes later, we ran in from the other direction. There was an immediate improvement in visibility, and the confident tones of Jock's corrections told me that he too was finding a crucial difference. This time he released our bombs, and with good results. With a stranger in the neighbourhood, I was glad to have completed the task.

Setting course for our return, it was already clear that a landing at Chittagong was on the cards. We had been airborne for eight and a half hours, and our remaining fuel would probably not last beyond another seven. It would depend on the winds we encountered, but reaching Salbani in one homeward leg seemed a slim possibility. Either way, it would be some five or six hours before I would need to decide which it was to be, and short of possible enemy action, it looked like being a pretty boring grind meantime.

For the first two to three hours, it was just that. Then things began to happen. Nothing dramatic, but gradually we had quite enough to attend to. First, an engine fuel pump began to malfunction. Red suspected it was icing up and that it might free itself if we could descend into warmer air. But with some distance to go before we were clear of the mountain ranges, there was no question of being able to free it off that way for the time being. Finally the pump gave up altogether, and I was obliged to feather that engine. This slowed our progress, but nothing worse.

Next we began to run short of oxygen, and as the flight proceeded I progressively removed from the supply those not directly involved in getting us safely to Chittagong: by now our immediate objective. As we started our descent only the navigator and the flight engineer remained connected: I had cut myself off a good hour earlier.

Approaching the airfield, Jack asked if I was going to have a shot at starting the dead engine. Having landed often enough on three engines without difficulty, I felt this was carrying caution a bit far, and continued the circuit on three engines. At about three to four hundred feet on a nice steady approach, Jack offered to select the landing lights. I seldom bothered with them, but since they could only make landing on a strange airfield that much easier, I accepted. (One of these devices was hinged into the undersurface of each wing and having been triggered, they took a few seconds to swing down into position.) Meantime, we were steadily losing height. As the powerful beams finally came into play, my view of the flarepath was suddenly obscured by a white glare as they bounced back off a veil of mist overhanging the approach.

The situation called for decisive action; to land or to overshoot. Although technically below the safe height for an overshoot, being temporarily blinded, I had no option but to do so. As Jack responded to my call for full power, the rudder force to maintain direction was suddenly beyond me and with the aircraft yawing strongly towards the dead engine, I yelled to Jack for his assistance on the rudder. For a few moments we slewed across the airfield in a most undignified manner, as we sorted things out and I built up speed for the climb. The aircraft being very light undoubtedly helped save our skins.

At a safe height, I took stock. Yes, I would attempt to restart the dead engine, but I would certainly not be using the landing lights on my next approach. I also concluded that, normal fatigue apart, my judgement and reactions were not as keen as usual, and that our shortage of oxygen had probably contributed to this. I informed the control tower that I was leaving the circuit temporarily.

With the dead engine successfully restarted, I handed control to Jack and sliding open my side window, took generous gulps of fresh air. I did this for a minute or so, then waited for Jack to give himself the same treatment before rejoining the circuit.

A few minutes later we touched down. But the gremlins had not quite done with us. As I gently eased the main wheels onto the runway – just to show that I had not really lost my touch – there was an almighty clatter; followed by an excited announcement from Ted Holmes, in the rear turret, that something was on fire at the tail. We had landed on a secondary runway built of steel planking, which slotted together like metal Lego; hence the clatter. The 'fire' was a shower of sparks from the tail skid, rushing past the rear turret as I rather overdid the aerodynamic braking.

What luxury it was to step onto the tarmac, and to stretch our legs. We had been airborne for fourteen hours and ten minutes: twelve of those hours at night. We still had a flight of an hour and three quarters to Salbani, but dawn would be breaking shortly, and on the next leg oxygen shortages and faulty engines would be mere memories. Meantime, we came down to earth in another sense, as one of the ground crew wished us 'Merry Christmas', and we turned to each other to exchange greetings.

Fortified by cups of hot char and bacon sandwiches, we were soon airborne and on course for base. How good the daylight seemed.

Even the tiger- and reptile-infested mangrove swamps of the Sunder-bans looked friendly. Debriefing at Salbani, always an affable experi-ence, was conducted with yet more cordiality than usual, as the intelligence officer handed round his cigarettes and, quite uniquely, produced an alcoholic drink before we set off for our beds. Like our sortie to Bangkok two months before, it had been a long night, and although lacking the operational drama, it had had its moments.

What lay behind the decision to attack that distant target, on what appeared to be a one-off basis, is anyone's guess. But the fact that a special-duties (SD) operations section had been formed at Head-quarters SAF earlier in the month, might not be unconnected: the role of the SD Section being 'to support clandestine activities in Burma, China, French Indo-China and Malaya'.

Direct Support for the Fourteenth Army

In the second week of January (1945) the priorities of the Strategic Air Force (SAF) were influenced by a new factor. The Fourteenth Army was approaching the river Irrawaddy along a front of over two hundred miles, and for their part the Japanese gave every indication of offering stubborn resistance along this defensive feature. With both sides aware that from early May the Irrawaddy would increase substantially in width and speed, the stage was set for an opposed crossing at an early date.

From this point until the end of February, SAF switched a good deal of effort from the long-range interdiction of strategic supply routes to operations in direct support of our army in the field. For my crew, this phase of direct support for the army opened on 13th January with an attack on Japanese barracks in Mandalay: not close support as such, but a softening-up of troops which might be called on to oppose our crossing of the Irrawaddy.

A number of SAF close-support sorties were flown during Janu-ary: our troops being only eight hundred yards from the target area in one instance. General Slim, the Fourteenth Army Commander, personally witnessed some of SAF's pattern-bombing attacks on close-support targets such as artillery positions: describing the results as 'magnificent' in his signals at the time. Later, in his personal account of the Burma campaign – *Defeat into Victory* – he again made generous reference to these SAF operations and since I was never near the receiving end of these attacks, and he was, I think his definition of them as 'Earthquake' bombing, can be taken as a fair description.

Aside from these direct support operations, January was something of a mixed bag for 231 Group. Among the gains: the standard of bombing accuracy was greatly improved following December's training phase; SAF squadrons, including 355 and 356, gave support to the invasion of Ramree Island off the Arakan coast; and airfields were successfully attacked – on occasion with P47 Thunderbolt fighter escort. On the debit side, 358 Liberator squadron, which had formed at Kolar as a heavy-bomber unit only two months before and had completed only one bombing attack, was removed from the Group Air Order of Battle on becoming a Special Duties squadron.

The Burma–Siam railway was also attacked during January. Nevertheless, the Group Operations record book for that month expresses concern over the shortage of precise intelligence on the whereabouts of the POW camps which provided the labour for the railway's maintenance, and there was a clear wish on the part of the air staff to establish a proper balance between the importance of severing this supply link, and to avoid, as far as possible, hazarding the prisoners. The apparent lack of a firm directive from higher command relating to attacks on the Burma–Siam railway added to the air staff's concern at this stage. For its own part, Headquarters 231 Group did its best to ameliorate the position by issuing its Wings and Headquarters 7th Bombardment Group USAAF with a map giving details of all known POW camps.

Speaking as an aircraft captain, I can say that when attacking the Burma–Siam railway we took great pains to bomb with accuracy, and invariably arranged our direction of attack so as to minimise damage outside the target itself. At the same time, as I've already indicated, our equipment and techniques fell short of being able to guarantee anything like complete accuracy.

January also caused the Group air staff some anxiety over the supply of Liberator aircrews. Up to that point, an operational tour of duty had consisted of three hundred operational flying hours. But the flow of crews from Canada had failed to match the timescale promised by the UK Air Ministry, and a conference chaired by the AOC met to find a solution. The outcome was a decision to fill the gap by increasing the tour of existing crew members to four hundred hours or one year, whichever came first: an added stipulation being that crews should not operate through more than one South-West monsoon period without a rest.

The impact of this decision to extend our operational tours did not reach 356 Squadron until March, by which time a number of us, including my own crew, were on the brink of tour expiry. For most, this was hard to stomach, and aircrew morale generally took quite a knock. For my own part I was able to accept the extension philosophically, but I recall more than one high-calibre person coming close to cracking up at the thought of his tour being extended: a flood of tears from a grown man can be very embarrassing. It was as though such people had subconsciously paced themselves to last a set number of hours, and an eleventh-hour change in the rules was the last straw.

Pathfinder Duties

To improve the accuracy of its night attacks, in early March 356 Squadron adopted the 'Pathfinder' technique, where selected crews went ahead of the main force to mark the target with incendiaries.

My crew was one of three or four to be chosen for these duties: a fact which pleased us, since it reflected confidence in our skill and dependability, and speaking for myself at least, because it added variety and individuality to the job. With some twenty-six successful missions to our credit we were in no way overawed by the task: on the contrary, we approached it with some relish.

We discharged two pathfinder tasks in quick succession. The first on the night of 2nd/3rd March, and the second on the night of 4th/5th March. The target on both occasions was the Makeson railway workshops at Bangkok, close on seven hours' flying from Salbani.

Our performance on the first raid was a textbook operation. We arrived within striking distance of the target with time in hand, and having loitered for a few minutes, we ran in precisely on time and placed our bombs squarely on target on the first run: all very satisfactory, but no more than I would have expected. Nevertheless, it gave me a certain elation as I banked steeply after our attack and looked back, to see our incendiaries burst into flame and illuminate the target. So much so that, on impulse, and knowing that the main force was not due to attack for several minutes, I circled the target area at a few hundred feet encouraging my gunners to silence the tracer fire from the ground.

I had found this combination of low flying and tangling with the defences quite exhilarating after the long outward flog. But it was not a clever thing to do, as a temporary but seasoned member of my crew was bold enough to tell me the following day. I must confess that, at the time, I thought him over-cautious and a bit of a wet blanket.

The second raid should have been a carbon copy of the first. Just as we had done two nights earlier, we arrived at the Initial Point with time in hand. And, once more, we loitered there so as to preserve the concentration of the attack. But having set course for the target, our navigation unaccountably deteriorated. Admittedly, the visibility that night was considerably reduced, so we had to work that much harder at pinpointing our position. And perhaps because we tried too much to map read our way to the target – instead of holding a steady course until the more distinctive features came into view – we got off track and behind schedule.

By the time we sighted the target, the main force was due to run in. Under these circumstances I dared not put my aircraft at risk to their high explosives by going in at five hundred feet as I had been briefed, and we were obliged instead to drop our incendiaries from two thousand feet. Although we hit the target, and made a contribution by stoking up the fires, at this stage the main force had already begun its attack. In the event, our fellow pathfinder crew had already marked the target successfully, so that our failure to do so on time was of no great consequence.

Nevertheless, I felt the most acute anguish at the mess we had made of things. To loiter just a few miles from the target as we had done, and then to come unstuck in the last ten minutes of a seven-hour outward flight, was agony indeed. Later, as we departed enemy airspace and began the long haul back across the Bay of Bengal, I felt so weary and fed up that I handed over control to Jack Clarke and flopped down alongside my dozing gunners behind the flight deck.

My hope was that I might sleep off my frustration. But it felt altogether wrong somehow being out of my seat, and I was soon back in the cockpit. It was the first occasion on which I had left my seat on an operational sortie, except to visit the Elsan. And it is a measure, perhaps, of how mentally shattered I was by our failure that night as pathfinders.

On our return, I recorded the outcome of our mission in the flight authorisation book, punctiliously noting the fact that we had only partially discharged our duty, as I bitterly recounted our late arrival on target to Eric Brighouse, my flight commander. Eric tried to console me, but I was in no mood to accept our failure for anything less than it was. If I was a bit strung up and over-reacting at that point, it is not too surprising perhaps: we had been on active operations for twenty-eight out of the preceding sixty-two hours.

Low-level Attacks

On 29th March I accompanied a new crew on their first operation, in which they were to attack bridge number 147 on the Burma-Siam railway. Our tactic on that occasion was a departure from the usual one of delivering our bombs singly in a succession of attacks, and involved dropping the lot in one stick, flying straight and level at five hundred feet.

Other aircraft had reached the target ahead of us, and the Japanese ground defences – already stirred up – were waiting. We ran in to an unusually hot reception: the air liberally spattered with the black blobs of exploding shells, as the AA gunners worked feverishly to get our range. But, not unusually, on that first run they did us no obvious damage.

As we completed our run and turned away, none of the crew seemed at all certain whether we had hit the target or not, and although our fixed camera might contain the answer, it would have been a bit frustrating heading for base without some idea of what we had achieved. So when the pilot I was screening asked me what I would do next, I said I would leave it to him, but that in his position I would make another run to establish our results. Having asked for my advice, maybe he felt under some obligation to act on it. Whatever his feelings that is what he did.

With one of the beam gunners briefed to take shots of the target with our hand-held camera, we made a second run. On this occasion the defences did not fumble. They found our range quickly and precisely. There was a loud bang as the perspex top of the front turret shattered and disappeared over the roof of the cockpit in a shower of fragments. Simultaneously, a piece of shrapnel came up through the cockpit floor between the other pilot and myself, leaving a neat hole in the roof: en route passing cleanly through a metal ration box and a map resting on top of it; otherwise barely disturbing either item. The latter unfriendly missile also shook months of dust from the cockpit carpet, which left us blinking for a minute or two.

Thankfully, we sustained no casualties. But the front turret was virtually unusable. The piece of shrapnel which had taken off the roof of the turret, had first struck the armoured glass shielding the gunner's face, leaving it completely crazed but otherwise intact. The other damage of significance was a punctured nose wheel. We were fortunate that the shell had not exploded a fraction of a second later, by which time it would have been inside the aircraft skin, instead of just outside. But such is luck.

Back at base, we landed on the secondary runway so as not to impede other landings, and by holding the nosewheel off the runway for as long as possible during the landing run, we were able to minimise the effects of the flat tyre.

The cost of our second run was remediable, although it would obviously keep our aircraft off operations for some days: a fact which Wing Commander Ned Sparks – who had assumed command of our squadron the month before – rammed down my throat when we met later in the Mess. What made him particularly annoyed was that the large majority of our aircraft on that operation had sustained damage of some sort, and it would be a few days before he could muster a worthwhile force.

Clinically, and with hindsight, Ned's objection to my tactics that day cannot be faulted. But, as I settled back in my chair and watched him retreat in a huff to his bungalow, I felt more offended than convinced by his instant admonishment. Taking instant decisions in the air is a bit different from negotiating a slam at bridge.

On 18th April, we turned our attention to the canal system west of Bangkok, successfully immobilising some of the lock gates. That day the squadron employed two methods of attack – both from low level, with a separate team of aircraft allocated to each method. One of these was modelled on our attacks on the Burma-Siam railway bridges which I described earlier (see pages 102 and 103). Whilst in the other – which my own crew used – we attempted deliberately to bounce the bomb into the target off a very low straight-and-level approach along the canal. Bombs are only too ready to bounce off water unless they enter near the vertical, and since ours were stowed horizontally, we had a head start.

With both teams busy in the same area, we naturally had to keep a close eye on what other members of the squadron were up to, and just as we delivered our last bomb, a yell from one of my gunners drew my attention to one of our aircraft attacking a target nearby. The aircraft, two hundred feet or so above its target, had pitched forward into a steep dive, leaving behind it an ominous puff of smoke. Within seconds, it had plunged almost vertically into the canal, impacting close to the lock gates it was attacking, and erupting in a ball of flame and thick black smoke. It was a shocking spectacle, and for a second or two I was numbed by the abruptness which which it had happened, since there had been no previous sign of ground defences.

Somehow, death on such a violent scale was doubly incongruous in that otherwise peaceful scene of lush green paddy fields, gentle flowing canals and pretty timbered houses. Not that our own activities were bringing peace to the neighbourhood in any immediate sense. To add to the poignancy of the occasion, my gunners, who had been watching the activities of the rest of the squadron in some detail, identified the crew as the one I had screened on the operation against Bridge 147, less than three weeks earlier.

What triggered that fatal sequence is a matter for conjecture, but I can advance one possibility. On that sortie, we had installed extra fuel tanks in the forward end of the bomb bay, where that initial puff of smoke appeared to come from. The possibility being that one of those tanks – at that stage empty of fuel, but containing a volatile mixture of petrol and air – had exploded on being hit by ground fire; which in turn had severed the elevator controls.

An Unexpected Announcement

One day in early April, word went round that the officers were to foregather in the Mess that evening for a special announcement, most of us receiving the news sometime towards midday. Over lunch, and at the flight offices that afternoon, speculation was rife as to what lay behind it.

With the Japanese on the retreat, were we about to be deployed further forward – even to the Pacific theatre of operations – in order to get within better striking distance of their main arteries of supply? Or with the impending German surrender, were we about to see the transfer of bomber command squadrons from the UK to bolster our attack? And if either of these scenarios was correct, why make the announcement in the officers' Mess and not in the Squadron Briefing Room where our NCO aircrew could also be present? Various ideas were put forward, only to be demolished as not adding up. Whatever it was, our superiors seemed unusually keen to achieve a hundred per cent turn out.

At the appointed hour Wing Commander Ned Sparks strode in and paused alongside me: close to the bar as it happened. As he took out his notes, before addressing the meeting, he remarked that he was glad to see me occupying a position of advantage. Whatever the announcement, it appeared that I personally was reasonably blameless. Indeed, I felt pleasantly relaxed as Ned stepped onto a chair and proceeded with his address.

He was reading a signal from Air Vice Marshal the Earl of Bandon, AOC RAF Bengal/Burma containing news of awards to members of the squadron. It began with the award of the Distinguished Flying Cross (DFC) to Squadron Leader War Harris, who had already departed the squadron on completion of his operational tour: news of which was extremely well received, for reasons which I hope my story has already made clear. The applause which greeted this news had barely died down, when I suddenly realised my own name was being read out, and that I too had been awarded the DFC.

The Mess seemed to erupt, as people pressed forward to offer their congratulations. And Ned promptly wrung my hand and bought me a drink. I was bowled over by the news. Certainly I had received no intimation of what was in store, and as I raised my glass to respond to all the well wishing, I found myself trembling with emotion.

Walter Stringer-Jones did his bit to keep my feet on the ground, by adding to his congratulations, the observation that whilst he would not say that he could think of no one more deserving of the award, he could not imagine anyone who would appreciate it more. A double-edged compliment if you like. Stringer was nothing if not a man of integrity. By then, however, I was too busy pushing the boat out to ponder his carefully chosen words. It was quite a party, and I remember finishing the evening in 355 Squadron's Mess: no doubt because they too were celebrating some awards.

A Double Watershed

The second of May 1945 saw the culmination ('D' day) of 'Operation Dracula': a combined land, sea and air operation aimed at liberating Rangoon, the Burmese capital.

The night before, as part of these proceedings, my crew carried out a special-duties operation in the general area: a mission designed to create a diversion. The SAF's main role in Operation Dracula was to assist in neutralising the Japanese defences on both sides of the Rangoon River, so that a brigade of the 26th Division could be put ashore on each bank and push northwards to the city.

In the event, however, after being put ashore from landing craft on 2nd May, the 26th Division was delayed by the sudden and untimely arrival of the South West monsoon, and an airman was the first outsider to enter Rangoon.

A JAPS GONE message had been spotted on the roof of the Rangoon central gaol by aerial reconnaissance on 1st May ('D' minus 1), and the following day when a Mosquito pilot of RAF 221 Group flew low over the city and could see no sign of the enemy, he landed at Mingaladon airfield – about eight miles north of the city – and walked into Rangoon. Having first visited the prisoners at the gaol and assured himself that the Japanese really had left the city, he acquired a sampan and sailed down the Rangoon River to meet our advancing troops. Incidentally, when he landed at Mingaladon airfield, this enterprising man had had the misfortune to damage his aircraft, but one likes to think that he did not have too much explaining to do on that account.

Even at the time, the capture of Rangoon was regarded as something of a watershed in the Burma campaign, and it was obvious to everyone that the Japanese were now, if not beaten, very much on the retreat.

My account of operation Dracula would be incomplete if I did not add the following, rather sad, tailpiece. As we left the dispersal, having completed our diversionary mission, we heard 355 Squadron start its engines for a follow-up raid. One of those aircraft was carrying Wing Commander Nicholson, the only Fighter Command VC of the war, who had attached himself from 231 Group Head-quarters, and had arranged to go along as an observer. He did not return. Following an engine fire, the aircraft ditched in the Bay of Bengal in the small hours of 'D'Day, and only two of the crew survived. Back at Salbani, Nicholson's small black and white dog spent the next few days wandering fretfully from one group of aircrew to the next, patiently seeking its master.

Within a few days of Operation Dracula, a second and more personal watershed occurred. I was told that my crew had done its quota of operational flying, and that we were free to take some end-of-tour leave. By this time, I had completed thirty-five sorties with 356 Squadron, and had exceeded the original operational tour of three hundred hours by twenty per cent. And yet, although conscious of being very fatigued, and in need of a rest, when I was given this news, I felt a strange reluctance, resentment almost, at having to stand down and hand over to others: not that I doubted their competence. Viewed dispassionately, that dichotomy of outlook may be difficult to reconcile. All I can say is that it was uncomfortably real at the time.

With that odd mixture of feelings, I visited the char wallah, and with a glass of sweet tea, went to a quiet corner of the crew room to make up my log book.

5   *Recuperation*

TB

With our log books submitted for an end-of-tour assessment, my crew and I set out for two weeks' leave at Darjeeling; a hill station some eight thousand feet above sea level on the borders of Sikkim and Nepal, and perched on the foothills of the Himalayas – the place that grows fine tea.

One reaches Darjeeling via its remarkable mountain railway–a mere two-foot gauge – an experience in itself, as the tiny steam train slowly puffs its way up the mountainside, using it would seem, every imaginable device to overcome the steep gradient. Not only is there an engine front and rear, and a man crouched over the front buffers occasionally dribbling sand onto the rails to increase the grip of the driving wheels, but here and there the lines are made to double back on themselves in an ingenious series of loops, S bends and figures of eight. Indeed there were times when we almost lost track of which was the front of the train; having entered a station heading in one direction and left it going in the other. Coupled with my experience of using the railway to the hill station at Ootacamund the year before, it was impossible not to feel a glow of pride in British engineering as we stepped off the train.

With the increase in altitude, it was not only the scenery and vegetation that had changed. The people too were different. By the time the train drew into Darjeeling, we had swopped the dark-skinned, and frequently querulous natives of Bengal for the stocky, good-natured Nepalis, with their paler complexions, their narrower eyes and their refreshingly childlike grins.

Aside from the local views in and around Darjeeling, which at first sight seem satisfying enough in themselves, if one really cares to make the effort there is a much richer feast to be had. That was why, a few mornings after our arrival, the more adventurous among us came to be standing on Tiger Hill – a local vantage point – as dawn broke. As daylight flooded the area, we had a breathtaking view of Mount Kanchenjunga, towering godlike to over 28,000 feet. It was so well defined, and exerted such a physical presence, that one felt it might be only ten miles away; whereas it was almost fifty, such was the clarity of that early morning air. On the same occasion we had a distant view of Mount Everest itself, as the sun's rays glinted on its eastern face, just over a hundred miles away.

As we drank in this magnificence, we knew that only half an hour later the intervening valleys would be filled with mist and cloud, and that the major features of the Himalayan range, now spread before us, would be blotted out, at least until dawn the next day. The reason for our getting up with the sparrows had nothing to do with bravado.

We began our leave with a full-blown party on the first night – not the most sensible thing to do perhaps, given that large increases in altitude temporarily exaggerate the effects of alcohol. But it is not every day that one successfully completes a tour of operations, so we felt good reason to celebrate.

But it did not stop at a party on the first night or even at one the night after. Speaking for myself, with the bracing tension of operations suddenly lifted, I felt jaded and on edge, and my remedy was to over-indulge in most things, with the exception of sleep. This reaction binge included dragging heavily on thirty-plus cigarettes a day, and it was not long before I had developed a hacking cough and had begun to feel distinctly below par. But in the circumstances, to cut down on the round of parties and late nights seemed unthinkable, unmanly somehow. Anyway, I was not deterred.

On our return to Salbani, my cough was so persistent that I was sent to the nearest Mobile Field Hospital (MFH) for a chest X-ray. Those in charge gave me the result on the spot. I had TB. 'There is a small cavity in one of your lungs,' said one of the two medical officers conducting the interview illustrating the size of the problem by comparing it with one of his fingernails. He proposed that I should be sent to Calcutta, where, he explained, that lung would probably be collapsed to assist the healing process. He went on to conjecture that I might be evacuated by boat quite soon, possibly stopping off in South Africa to help speed my recovery. As I left, the two doctors further dramatised the moment by getting to their feet and shaking me sympathetically by the hand.

Taking stock of my position, I barely registered the return journey to Salbani. I resolved that before writing home, I would first wait for the hospital authorities in Calcutta to give me their prognosis. Not that I doubted the basic diagnosis. My Uncle Burt and Auntie Winnie had both died of tuberculosis, and I could imagine the consternation my news was going to cause. But, if I could sketch out the plan for my recovery at the same time as breaking the news of my illness, it might at least add some reassurance for the longer term.

That night I packed my belongings, and the next day I was on a train to Calcutta. It was a miserable farewell. Most of my close friends in the Mess were founder members of the squadron, and had recently been posted on completion of their tours of duty. And although I had struck up some budding friendships among the replacement crews, these had yet to achieve real depth. My crew members were as stunned as I had been, but when you've picked up an infectious disease – whilst you receive heartfelt expressions of sympathy – no one wants to get too close to you. I felt pretty low at that moment.

At the Military Hospital in Calcutta I was promptly examined by Wing Commander Robson, an eminent chest specialist. As he completed his examination and scanned my medical report, he said, 'Well old son, you're pretty worn out. But give us two or three weeks, and I reckon we'll just about have you right.' I challenged his extraordinary optimism, and asked how he squared that with the diagnosis delivered to me verbally at the field hospital. I suggested he might have missed something in the report. He had. As he turned over a page, he appeared for the first time to see the crucial part.

The consultation snapped into higher gear. The wing commander immediately assured me that nothing in his examination suggested that things were as serious as I had been told at the MFH. He would have me X-rayed forthwith, and would personally inform me of the results as soon as they were available. He gave me additional encouragment by explaining that mobile field hospitals were not equipped with a certain piece of apparatus necessary for the definitive interpretation of X-rays. Why then, I wondered, had those MFH doctors seen fit to pronounce so positively on such a serious matter? I decided that it might be a little premature to open the champagne, but I slept better that night.

Early the following day the Wing Commander put my mind completely at rest. Not a trace of TB; although a scar or two indicated an earlier infection of some sort. What a relief! And how fortunate that I had decided to wait before writing home.

My personal anxieties lifted, I looked more closely at the other people in the ward. In the bed opposite, Wing Commander Hill, his emaciated body covered in ugly sores, was recovering from years of Japanese imprisonment in the Changi gaol in Singapore. Apart from being on a special diet, reinforced with vitamin tablets, he was occasionally turned onto his side to receive an injection: penicillin perhaps. One sensed his recovery would be a long haul. My own ailment narrowed into perspective.

Within a day or two I had a visitor from Salbani. Flight Lieutenant Les Fayle, the captain of one of the newly joined RCAF crews, had made the journey primarily to settle a gambling debt: the equivalent of £40 pounds and not chicken feed in 1945. As far as Les knew before his arrival, I was likely to be invalided out of the country at any time, and his determination to settle up before that could happen impressed me no end. It was nice too, to be able to inform someone from Salbani in person that the earlier diagnosis had been over-turned. As for my feeling badly about Les's loss, well it could equally have happened to me, and since he was responsible for introducing craps (the game in question) into the Mess, that was something I could live with.

It was then the beginning of June, and it took two to three weeks of hospital treatment, as the chest specialist had predicted, before I was released.

With a railway warrant and a chit for two weeks' sick leave, I made my way to Puri on the Bay of Bengal, booking in at the Bengal-Nagpur Railway hotel. Long established, and a favourite with Europeans, the hotel exuded an atmosphere of calm efficiency. The rhythmic breaking of the tide, clearly audible from my room, adding to the hotel's soothing ways. The food too was good; delectable curries and freshly caught fish being the specialities.

Peaceful though it was, the hotel provided me with some convivial company in the guise of a Scots oil engineer and an Indian Army major. The oil engineer had been booted out of Burma by the Japanese in 1942, and was waiting in the wings to make his return. The major, like me, was taking a rest. As a rule I met them over a drink, so I knew them best in that context.

Overtly, at any rate, the major began his drinking day at eleven in the morning with claret. He was a man of habit, arriving punctually on the verandah at that hour bearing a drinks basket, the latter handcrafted to contain two bottles and a supply of biscuits. Towards lunch, he switched to large pink gins. With his neat moustache, crisply laundered bush jacket and gold-rimmed eye glass, he looked frightfully pukka – a trifle studied perhaps, with the claret and all that, but very sociable none the less.

The Scot on the other hand was more homespun, and had an altogether easier, more down-to-earth manner. He had a Tennants lager at lunchtime, and left the spirits until the evening, when he was happy enough to indulge in a wee dram or two. Given my convalescent state, it was his prescription I adopted.

A day or so of peace and calm found me reflecting on the events of the past year, and I took the opportunity of writing a short letter to each of my crew saying the sort of things which I might have said at the time I left Salbani had the circumstances of my departure been less traumatic. With time to think, I was conscious of having made heavy demands on them during our tour of operations, and wanted to thank them for the way they had responded, and to wish them well for the future.

As my energy returned, there was no shortage of things outside the hotel to engage my interest. Nearest to hand were the activities of the fishermen, who operated from the beach within a few minutes' walk. They were short, wiry men, with ragged clothes and the serious faces of those who have to graft for a living. The state of their clothes was reflected in the outward condition of their boats: frail looking craft which had seen years of service, and which were liberally patched, and lashed together in places with bits of rope. But if there was an appearance of make-do-and-mend about these men on shore, there was nothing unprofessional about the way they handled their boats. In that capacity they worked impressively as a team, with only an occasional word passing between them.

The boats were launched, as one might expect, bow first, with the crew running into the sea until they were practically up to their armpits, as they pushed their craft out before heaving themselves aboard. Then it was a case of maintaining the seawards momentum with crudely fashioned oars until the boat's sail filled out. Accompanying one of the crews, I was so busy taking in all this skill and expertise, that it was a while before I realised there wasn't a sign of a lifejacket on board: not even one. It was at that point that I had began to have reservations about the completeness of their professionalism. Half a mile out to sea with the boat beginning to creak with the freshening wind, I was not at all disappointed when we headed back with our catch.

Beaching the boat was done stern first, and called for expert timing. We loitered about thirty yards out, oars at the ready. When a suitably large wave came along, the boat was smartly put about and the wave picked us up and shot us onto the beach. Usually, this operation went like clockwork, but if it was mistimed and the boat was left short, the undertow promptly sucked it back into the water, and the process began again.

The undertow, which varied with the tide, was something which would-be swimmers too had to take account of. At times, exceptional swimmers apart maybe, it was best not to venture in at all. But, regardless of the tide, the beach was a pleasant spot in its own right, and I gathered from the Scots engineer that it was even more popular with the residents in pre-war days. Apparently, with its full peacetime complement of staff, the hotel could afford to station a drinks waiter on the beach as a matter of routine. And, as if that was not enough, so that the customers were not kept waiting, he signalled their requirements to another member of staff hovering between him and the hotel. And I thought *I* was living the life of Riley!

Puri did not depend solely on its fishermen for its livelihood. Far from it. It was a major centre of pilgrimage for Hindus. Every day its streets were crowded with pilgrims making their way to and from the large Hindu temple in the centre of the town. And where there are people, there is money to be made: whether it is in supplying refreshments, souvenirs or, as in this case, the literature and knick knacks of religion. There was not much the traders of Puri had to learn about milking the visitor.

The constant stream of visitors gave the town a vibrant, holiday atmosphere, and visiting its bustling streets it was impossible not to feel uplifted by the colour and vigour of it all. There was more than a handful of holy men among the visitors it's true, with their long white beards, their ascetic faces daubed with white stripes, their piercing eyes and their scantily clad bodies; but in the main the visitors comprised ordinary Indian families – every member scrupulously turned out for the occasion in their best clothes, the little girls sporting marigolds in their glossy hair, and all radiating the happiness and anticipation of a day out.

The particular Hindu diety celebrated by the Puri pilgrims is Krishna; variously described as slayer of demons, flute player and lover. Once a year, scores of his followers drag his idol through the streets of Puri on a huge eighteen-wheeled wooden car. For eleven months of the year, like some dead dinosaur, this vehicle stands idle, other than to arouse the curiosity of the pilgrims and the attention of an occasional western visitor like myself. But in the weeks leading up to the event, it becomes the focal point of activity as its lofty superstructure is decked out in the colourful materials befitting a festival.

In a curious way one of Krishna's titles, 'Lord of the World', has also found its way into our own everyday language. In Hindi that title is expressed by the single word *Juggernaut* (or *Jagannath*), and since that word has become – in British minds at least – as much associated with the huge multi-wheeled Puri car as with Krishna himself, we have adopted it to describe large multi-wheeled vehicles in general, although usually in a derisory sense.

Personally, I did not venture into the temple, but from souvenirs on sale in shops nearby, one got a fair idea of what lay within. It came as no surprise that the Hindu gods – especially Krishna – were well represented among these souvenirs, which for the most part consisted of small soapstone carvings, accompanied by effigies in enamelled brasswear. But what did surprise me was the presence of some quite explicit erotic carvings among the reproductions on sale. Admittedly, in the shop I visited, the latter were discreetly housed in a small curtained-off section of the shop – presumably so as not to offend those with young families – but the proprietor assured me that these were faithful reproductions of what was inside the temple. Later I came to realise that eroticism is well established in Indian art. But this first encounter, arising as it did in the context of a holy shrine, came as a real shock.

A Taste of Administration
On my return to Calcutta I was given a complete aircrew medical examination, passing all tests but one. When it came to the colour vision test, the medic's eyebrows fairly shot up as my deficiency was revealed. I explained that I had entered the RAF with a colour vision grading of 'Defective/Safe'; to which he retorted, 'Defective – yes, safe – no.'

Before getting my wings, and for a while afterwards perhaps, I would have been shattered by such an announcement. But now that I was an established pilot, I was able to receive this uncompromising pronouncement without feeling the least bit threatened. I pointed out – a little smugly perhaps – that I had recently completed a successful tour of operations, which had included over one hundred hours at night, and I suggested that that fact could hardly be ignored. The medic grudgingly admitted that I had a point, and scribbled a proviso to the otherwise damning results of my test. I was left in no doubt, however, that if that old codger had conducted my recruitment medical, my aircrew aspirations would have been torpedoed there and then.

Pending the arrangement of an aircrew posting, I was given a temporary job running a Calcutta transit camp. My particular unit was a self-contained off-shoot of a parent site, which meant I had a large measure of autonomy: a situation much to my liking. We handled RAF and Fourteenth Army personnel – mostly the latter – returning from Burma after the expulsion of the Japanese. Many of these men had served a full three years in the theatre and were staging through on repatriation to the UK. In the case of the soldiers, years of jungle fighting had left its mark, physically and mentally. Most of them were gaunt, hollow-eyed and extremely tense. Their one abiding anxiety – to be reunited with their families.

One such man sought an interview with me. In a lather of anxiety, he first asked for my assurance that whatever he might disclose, it would in no way delay his repatriation. I explained that I could not go that far, but I agreed to give his request every possible consideration. In turn I asked for his understanding of my duty, and also his, to balance his personal interests against 'the wider interests of the Service'. As I uttered that trite phrase, and watched his tortured expression, I was sharply reminded of the yawning gap which occasionally separates the individual serviceman's view of his priorities from that of his unit commander. I waited in silence for some seconds, before he conceded my point and came out with his story. He had accepted a small package from a stranger in Rangoon, with the request to post it on arrival in Calcutta. En route, he became suspicious of its contents, and on opening it found it contained gold.

This was clearly not some local matter which I could keep the lid on, and on the basis that the sooner the matter was investigated, the sooner it would be resolved, the soldier agreed to be interviewed by the service police. My contribution was to stress to the police that our informant had come forward voluntarily, and to ask that their enquiry should interfere as little as possible with his return to the UK.

The bulk of my time, however, was not concerned with the individual problems of the transit population. Rather, it had to do with directing and co-ordinating the provision of accommodation, meals, spiritual welfare, transport and general administrative support. To assist me, I had three excellent senior NCOs and a handful of ubiquitous corporals and airmen.

It was my first experience as a full-time administrator, and apart from keeping a diary as the basis for follow-up action, I relied heavily on my staff. I consulted them frequently, and they responded with enthusiasm and know-how. In fact there were times when, in his eagerness to get the job done, one of the senior NCOs imprudently ignored the limits of his own area of responsibility, and I found myself having to rein him in in order not to upset the more staid members of the team. But in spite of such hiccups, within a fortnight I felt I had some measure of the job, and almost enjoyed the challenge of accommodating close on a thousand personnel: often at the drop of a hat.

However, I had not joined the RAF to become an administrator, and at the beginning of August, I was glad to receive instructions to resume flying duties. I was to join the staff of the Liberator Refresher Flying Unit at RAF Kolar in mid-month.

What I had not bargained for was the fuss my superior made about losing my services. Not surprisingly, his protests were shrugged off by higher authority. But having lost his case, he did not immediately abandon interest in me. Instead, he found time to organise a farewell party, making it an all-ranks affair, so that my staff could attend. Apart from being quite flattering, it was a generous act, which gave both my morale and my respect for authority quite a boost.

About a week before I departed for Kolar, the newspapers reported the dropping of an atomic bomb on Hiroshima. It was described as a weapon of immense power which was expected to shorten the war against Japan dramatically. A few days later news came of the dropping of a second atomic bomb. Suddenly, what had looked like being a long and bloody struggle, was in prospect of rapid termination.

Among those who would have had to do the actual fighting if this were not so, there was considerable relief, as well as a good deal of satisfaction that a brutal and arrogant enemy would shortly be defeated.

6   *After Hiroshima*

Adjusting To Peace

The Liberator Refresher Flying Unit (LRFU) at Kolar was set up to restore the edge to trained Liberator crews before they joined their operational squadrons, dealing in the main with crews newly arrived from the Boundary Bay conversion unit in Western Canada. Classically the LRFU did its job in two stages: first, by refreshing crew members in the basic handling of the aircraft and its equipment, and afterwards polishing up their operational techniques across the board, from locating and bombing the target to defensive tactics against fighters.

My posting was to the Operations Flight. But by the time I arrived, even the lowliest erk must have begun to question the relevance of its function – if not the future of the LRFU itself – given developments in the political field. (In the week after the dropping of the second atomic bomb, the Japanese emperor publicly announced Japan's capitulation.)

Naturally, in this state of affairs, the least responsible thing would have been to put our feet up and to have waited for 'the boat' to take us home. So, for the time being – other than practice bombing and fighter-affiliation exercises taking a back seat – it was business as usual. Before I personally could play my part as an instructor, however, there was the need to get back into flying practice. My first two weeks at the LRFU were spent on a series of navigation exercises; visiting, among other places, China Bay on the east coast of Ceylon.

That particular visit provided two unusual sights: first, the formidable shape of the French battleship *Richelieu* which lay at anchor under our approach path, and whose massive bulk contributed to the turbulence as thermal currents rose from its massive bulk in the midday sun; and second – and in its way more dramatic – the twin fins of a Liberator (from a Dutch maritime squadron) protruding above the surface a few hundred yards offshore. The latter providing a grim reminder of how critical it was to keep the Liberator's centre of gravity within limits when taking off with a full load.

In the fortnight that followed, I flew with a number of 'trainee' crews: my role being confined to checking their basic drills and procedures. In the things that mattered I found no black sheep. And as to minor discrepancies, whilst there was no question of encouraging sloppiness, it was not the easiest thing during that slightly bogus period to get worked up over the finer points.

With job satisfaction on the wain, I was delighted when – in late September – the station commander approached me with a job which had some relevance to the post-war situation. (The official Japanese surrender had taken place on 2nd September, on board an American aircraft carrier in Tokyo Bay.) It sounded a plum job which Group Captain Le Poer Trench was offering me, and I agreed to it without hesitation. I was to take a Liberator and put it and my crew at the disposal of the Recovered Allied Prisoners of War and Internees (RAPWI) organisation at Bombay. On asking for my terms of reference beyond that, the Group Captain – who was not a man for detail – said, 'Oh the RAAPWI chap will fill you in on that; just take him where he asks.'

A day or so later, having reported to Major Henderson of RAPWI, we were poised for action at Santa Cruz airfield. But after a week of lolling around the Mess, I began to wonder what had happened to the initial sense of urgency which had brought us hurrying across from Kolar. Although the Santa Cruz Mess had an unusually good supply of beer and a sandy beach nearby, I felt these to be poor substitutes for action: a view fully shared by my young 'student' crew.

Finally the Major did get his act together. On 8th October we took off with several sacks of mail for delivery to the ex-POWs and internees then staging through Rangoon. We night-stopped at Dum Dum on the outskirts of Calcutta, and the following day headed south-east across the mouths of the Ganges and out over the Bay of Bengal. How familiar it all seemed, and yet how strange not to be testing our guns and moving into a higher mental gear before entering Burmese airspace.

It was late afternoon when we landed at Pegu, a steel matting strip about fifty miles from Rangoon, and by the time we had transferred the mail into a thirty-hundredweight truck and secured the aircraft for the night, it was already dark.

I still tense up when I recollect that road journey to Rangoon. Major Henderson sat alongside the driver, whilst I sat a few feet behind, perched uncomfortably on the sacks of mail in the back. The driver was a Sikh soldier in his late teens; fresh from his drivers' course I suspect. Certainly he was still driving by numbers. The road was badly rutted, which did nothing to help the situation, but the real problem was our driver's complete lack of anticipation when negotiating the three hundred and fifty bridges between Pegu and Rangoon: it seemed like three hundred and fifty anyway.

To save material, the bridges were considerably narrower than the road itself. Which suggested to me that, on the dark night in question, we should either be hogging the crown of the road, or at least steering for it at the first sign of a bridge. Our young driver however – a very law-abiding soul it seemed – approached every bridge well over to the near side. And since he appeared consistently to see the bridge at the very last moment, every crossing became a heart-pounding adventure as he swerved violently to avoid colliding with the left-hand upright. Peering ahead between the major and the driver, I had a ringside seat; and although I could hardly bear to look, neither could I turn away and rest easy. By the time we reached Rangoon I was in need of a strong drink.

During our short stay I took the opportunity of visiting the Schwe Dagon Pagoda: an imposing landmark with its lofty golden spire, and one which I had seen once or twice from the air in less peaceful circumstances. At the entrance, I removed my shoes at the invitation of a minor official; the latter betraying not the slightest hint of

curiosity that these days his visitors were European rather than
Japanese. His stoicism seemed at one with the general air of permanence radiating not only from the sheer size and elegance of the
pagoda itself, but also from the little acts of conformity being played
out by the monks and novices in their bright saffron robes. Commer-
cialism too had found its niche, which it somehow managed to
occupy without appearing at all mercenary: on sale in the outer
courtyard were bunches of sweet-smelling jasmine to place as offer-
ings at the small individual shrines, as well as wafer-thin slivers of
gold leaf to donate towards the regilding of the spire.

By way of physical recreation, I was persuaded to take a night-
time swim in Rangoon's Victoria Lake. But in spite of the warm
water and the pleasantly cool evening air, there was something
vaguely menacing about bathing where the Japanese had so recently
been in occupation, and I was quickly in and out. As I went through
my brief repertoire of strokes, I recollect hoping that the Japanese
had not stocked the lake with crocodiles as a parting gesture. If
they had, perhaps it was the brevity of our swim which spared us
their attentions.

We returned to Pegu in daylight, thirty-six hours after our arrival.
How different from the nerve-racking inward journey. The aircraft
had remained serviceable, and after breaking our journey that night
at Calcutta, we were back at Santa Cruz airfield the following afternoon. We spent a further week at Bombay, but no further RAPWI
business came our way, and on 19th October we returned to Kolar.
(Months later, I learned that the air staff in Delhi were mystified by
reports that one of their Liberators had landed in Burma, since they
had authorised no such movement. Well, what would life be without
its surprises? On our return to Kolar, the good Group Captain Le
Poer Trench had merely been interested in whether we had done a
useful job and enjoyed the assignment. Do those two reactions, I
wonder, characterise the difference between bureaucracy and
pragmatism?)

By this time, the United States government had caught up with
its housekeeping *vis-à-vis* the military equipment it had leased and
lent to Britain. Effectively our Liberators were already grounded,

and a month later we began to ferry them to the RAF maintenance unit at Cawnpore, to await disposal by the American authorities.

The sad task of handing back our beloved 'Libs' was completed in early December, by which time Cawnpore airfield, having also absorbed the Liberators from the operational squadrons, had assumed the depressing atmosphere of a graveyard. Rows of Liberators, wing-tip to wing-tip, flanked one of the runways: the irony being that, although most of the aircraft were in excellent condition, we knew that few, if any, would see useful service again.

We returned from the last ferry sortie on a routine communications flight from Delhi; an arrangement preceded by a chilly overnight rail journey from Cawnpore. However, it seemed that we would gain some compensation from the sad business of parting with our last aircraft. Only forty miles off track, at Agra, lay the Taj Mahal, the fabulous white marble mausoleum built by the Moghul Emperor Shah Jehan for his wife Mumtaz-i-Mahal. Our flight crew were more than willing to make that small diversion, so that we could have a brief glimpse of it. But arriving overhead, we found that Agra and its surrounds were still enveloped in early morning mist. There was no alternative but to resume our game of liar dice, and to hope for another opportunity.

A pleasant surprise awaited me on my return to Kolar. It was a personal letter from the Sales Director of Fox Brothers, the woollen cloth manufacturers in my hometown, tentatively offering me the post of assistant to their London representative. I was invited to take my time in thinking it over. It was a handsome gesture on the part of Harry Fox and, with my demobilisation probably less than a year away, a reassuring one too.

A Sticky Beginning As a PA

A day or two later, I was asked if I would like to be considered for the post of Personal Assistant (PA) to Air Commodore Waring, the Air Officer Commanding (AOC) RAF 225 Group. I was flattered at the thought, and agreed to be interviewed. The AOC's area of command comprised the whole of Southern India; his northern boundary taking in Bombay on the west coast and Vizagapatam on the Bay of Bengal. Group Headquarters were pleasantly situated overlooking the golf course in Bangalore, a garden city thirty miles west-south-west of Kolar.

Entering the AOC's office, I was confronted by a small, balding man. His mood was distinctly tetchy, and I wondered what I could be letting myself in for. Almost his first question was whether I planned to make the Service my career. With unnecessary candour perhaps, I revealed my job prospects outside the Service and, in doing so, implied that I would be leaving when my age-and-service group was released. What I did not say was that, given my lack of scholastic achievement, I felt I had no option but to seek my future outside.

The Air Commodore was clearly put out by my response. He saw the PA's job as a means of providing valuable experience for someone aspiring to a full career. He also had some difficulty in reconciling my wanting the job with the fact that I was not hell-bent on remaining in the Service. Nevertheless, I was accepted and immediately took up post.

Air Commodore Edmund Francis Waring CBE DFC AFC, known to his friends as 'Tony', had been awarded his Distinguished Flying Cross for the destruction of a German Zeppelin during the First World War. His CBE was a more recent honour, marking his successful term as director of air-sea rescue policy at the Air Ministry. Under his guidance a lifeboat was developed which could be dropped from the air (the Lindholme Dinghy), greatly enhancing the possibility of rescuing crews which had been forced to ditch.

He had begun his service career with the Royal Naval Air Service, and much of his time in the RAF had been spent on flying boats. Apart from the obvious advantages of his maritime background, I imagine his practical grasp of engineering problems played an important part in his success with the airborne lifeboat, and I was to see more of this practical trait in his handling of problems in the months to come. Privately, he was fond of saying that he made no claim to be clever, but that he could draw on a wealth of experience.

He was essentially a shy man, placing great store on personal integrity and loyalty. And if he felt he had been taken advantage of, he could be quite upset. The underlying reason for his prickly mood at my interview was that his chief, Air Marshal Sir Roderick Carr AOC-in-C India, had robbed him of his previous PA virtually overnight: a fact I discovered later.

Losing a good assistant at no notice was bad enough. But what added to the AOC's irritation, was the thought of having to build up once more that special trust which must exist between senior commanders and their close personal staff. For a shy man this would be doubly taxing. In fact, we settled into a very good relationship. But not before a rather sticky incident, which could conceivably have cost me the job.

This occurred during a visit to Bombay. I had been in post a mere three days, and the AOC and I had still to build up a rapport. The outward journey in his personal aircraft – an American Beechcraft Expeditor – went smoothly enough; the car I had arranged was waiting at the airfield to take us into the city, and the Royal Bombay Yacht Club, where we spent the night, greeted the air commodore with every courtesy. Dinner too, which we ate at the club, and which was hosted by Air Commodore Peter Craycroft the local RAF base commander, went like clockwork. It was after dinner that my difficulties began.

Over the brandy, the base commander told me of a change to the AOC's itinerary necessitating a delay in his arrival at Santa Cruz airfield the following morning. Before we turned in, I did my best to get this across to the AOC; emphasising the point that our departure from the Yacht Club would be delayed as a result. But he was in no mood to talk business, and having done my best, and been told to stop fussing, I left him to get some sleep.

I was still shaving when the AOC burst into my room the next morning and roundly choked me off for being late. He spared me nothing, including a reminder that I was meant to be looking after him, whereas it seemed that he was having to wet nurse me. Clearly, he had not taken in my message of the night before about our delayed departure, and although I tried to put him right, he cut me off and stalked out. I completed my ablutions and joined him for the journey to the airfield. By then, I had concluded that, although it clashed with my instincts of loyalty, he would have to learn the hard way that I could not assist him if he persisted in shutting me up. We motored the forty minutes to Santa Cruz in silence.

Our early arrival caused no little embarrassment to the station commander and, once the explanations had been heard, to the AOC also. It did not, however, clear the air between the AOC and myself. On the three-hour flight to Bangalore, he remained withdrawn.

Fortunately, at Bangalore a fault with the aircraft came to my assistance. Our pilot could not lower the undercarriage, and since a belly landing seemed in prospect, I asked the Air Commodore if I could speak to the pilot, since I might have a solution. He snapped back words, to the effect that if I had a solution, of course I could: emphasising his reply with some invective.

Within seconds of my arrival at the cockpit, the undercarriage was down. Thanks to my familiarity with American aircraft systems, I had guessed correctly what the fault might be – an electrical circuit breaker which had tripped during an overloading of the system. As soon as it was reset, the wheels came down with a satisfying 'clunk'. Our pilot was obviously new to the problem.

Overcoming that emergency lanced the boil. On the journey to his residence, the Air Commodore was gracious enough to admit that he had, perhaps, been a bit hasty earlier in the day, and suggested that I might dine with him that evening. From that moment I had gained his trust.

In less than a fortnight we were into the Christmas festivities, with the AOC attending the serving of the airmen's Christmas dinner, and on Boxing Day presenting the prizes at the Group headquarters sports day. But, for me, *the* official event of the Christmas break was accompanying him to lunch at the Madras Sappers and Miners Officers' Mess, where we were hosted by his younger brother, Colonel Roscoe Waring.

The Madras Sappers and Miners was no ordinary regiment, being first raised in 1780. Madras, as my reader may know, was the birthplace of the Indian Army, whose origins go back to 1639. And one of the reasons why the Army of Madras remained loyal during the

1857 mutiny has been attributed to its sheer longevity. Whatever the reason, that loyalty gave such regiments as the Madras Sappers and Miners a place of special regard in British political and military circles.

We first sat on the Mess verandah and, by way of an appetiser, downed a dozen oysters apiece; washing them down with Guinness or a pink gin according to one's taste, my chief coming out of these preliminaries rather impressively by plucking a pearl from his mouth at one stage. The Sappers and Miners did not do things by halves: the oysters were brought in on huge silver trays, and everyone having agreed to the addition of a little red pepper, a large wooden pepper pot was shaken double handed over each tray in turn.

Later we moved to the dining room and, having disposed of the Lady Curzon soup and melba toast, we set about a superb curry, served with crisp poppadams, chapattis and dish after dish of exotic additives: Bombay duck (dried fish), dessicated coconut, mango chutney and a mixture of chopped onion and chillies, to mention but a few. All this in an atmosphere rich with permanence and tradition, characterised by the battle honours on the walls, the silver trophies and mementoes which decorated the table, and the attentive waiters in their handsome uniforms. The Army's hospitality rating took quite a leap that day.

With the celebration of the new year a few days later, one might have expected the social scene to peter out for a while. But this was not to be. Under the sponsorship of the AOC, a number of his staff became involved in arranging a charity ball in aid of St Dunstans. And since everyone of importance in Bangalore, from the British Resident (senior diplomatic representative) downwards would be attending, nothing, but nothing, was left to chance: be it protocol, organisation, publicity, or whatever.

Coping with this fund-raising job in addition to progressing their day-to-day duties, gave a number of the AOC's staff more than enough stimulation over the two to three weeks' preparatory period. Thankfully, in spite of last-minute alarms backstage, the event went off without a hitch, and more important, it raised a handsome sum.

Among the organisers, however, there were few regrets at seeing the task successfully behind them.

A Turbulent Peace

Although the war with Japan was over, HQ 225 Group soon discovered that 1946 was going to be anything but peaceful in the problems it posed. At that stage, the British forces in India still had a degree of responsibility for rehabilitating the countries recently under Japanese occupation. And plans previously formulated for the stage-by-stage freeing of these countries were being drastically reshaped in the wake of Japan's abrupt collapse. With a number of the front-line squadrons concerned in this reshaping lodging at 225 Group airfields (at Bangalore and Madras for instance), their commanders became frequent visitors to our headquarters.

But it was not long before the bulk of our problems stemmed from a quite different and unexpected quarter. Repatriation and release from the service rapidly became an obsession with a number of the RAF conscripted airmen, who felt that authority was moving far too slowly in the matter. They reasoned, for example, that bomber aircraft and crews 'now standing idle', could easily be adapted to the transport role and used, together with proper transport aircraft, to augment the troopships.

This mood became quite vocal on some of our stations. Unofficial meetings were called by a vociferous minority of airmen, who urged their colleagues to withhold their labour unless promises of improvement were forthcoming. Such meetings were sometimes addressed from the back of the hall, so that the anonymity of the ringleaders was preserved, and action against them made that much more difficult. Evidence also suggested that the signals network was being used in an attempt to spread the threat of the 'strike' weapon.

The Judge Advocate General's representative was a frequent visitor to the AOC's office – and to his bungalow for that matter – as the legal·implications of these 'strikes' were considered. And not surprisingly, there were periods when we were deluged with queries from Air Headquarters Delhi, our senior formation, anxious to keep abreast of developments in this politically sensitive issue.

There was nothing that could be done at the local level to speed up repatriation and release in the immediate sense, but we could at least attempt an accurate assessment of how general the mood of dissatisfaction was and to sustain morale by ensuring that complaints about local issues were dealt with effectively. To this end, the AOC visited each of the stations affected, addressing the airmen and making a personal assessment of the problem. Returning from these visits he was invariably locked in discussion with his staff for hours afterwards, as he sought answers and solutions.

News of the 'strikes' soon reached the politicians at home, and it was not long before Air Marshal Sir John Slessor, the Air Council Member for Personnel, visited India to satisfy himself that everything possible was being done to preserve morale and discipline. My AOC flew to Bombay to receive him, and as one of those present at a short speech which Sir John made on his arrival, I was most impressed with the matter-of-fact way in which he dealt with our difficulties. His calm pragmatism certainly helped me to retain a sense of proportion over the way in which our traditional values of discipline and behaviour were suddenly being challenged. I had noticed too how my own chief's vast experience had occasionally served to steady members of his staff in the handling of the 'strike' problem in the weeks before Sir John's visit.

Our own forces were not alone in airing their dissatisfaction during 1946. The crew of an Indian Navy vessel in Bombay harbour became mutinous, making various demands of their authorities on shore. The vessel was armed, and implicit in the mutineers' demands was the threat of force to resist the vessel being boarded.

In taking stock of their options, the naval authorities contacted our headquarters to see what assistance we might provide if called upon. One possibility considered was to have a ground attack squadron fly past at low level, to demonstrate that authority too had teeth it could use if it was obliged to.

To the best of my recollection this contingency was not invoked, but I remember it giving my AOC and his air staff cause for much head scratching at the time. Our RAF rocket-firing Mosquitoes, based at Madras, had been temporarily grounded because the resin-bonded spars of these all-wooden aircraft had been weakened by the high ground temperatures. This meant that if these aircraft were used, the pilots would be limited to gentle manoeuvres, not really suited to a task which could conceivably develop beyond a simple fly-past.

The alternative of earmarking one of the Indian Air Force squadrons for the job was also considered. But the problem seen with this solution was that it might impose unacceptable strains on the loyalties of the personnel involved had force been ultimately necessary. Choosing the right alternative in this delicate situation was another of the conundrums which saved my chief and his air staff from vegetating in the post-war period.

The first half of 1946 also brought problems for the native population of Southern India, when a serious famine struck the state of Mysore. The United States, once again playing the role of benefactor, reacted by sending former President Herbert Hoover on a fact-finding mission. Waiting to greet him on his arrival at Yelahanka airfield was the Viceroy (Lord Wavell) and the Maharaja of Mysore: having themselves already been welcomed by the Bangalore Resident and the local Service chiefs. We junior aides also stood in line, just in case there was a fast ball to be fielded. But as so often happens with the really big fish, no show crises were staged. And after a few minutes' quiet chat, they got into their cars and headed off towards the city of Mysore and the heart of the famine.

The job of personal assistant contained plenty of variety, and it seemed to me that fatigue only set in during the occasional quiet period.

Off To the Hills

In contrast with, say, Delhi, which obeys the more general rules of the Indian seasons, Bangalore has a miniature climate of its own, being on the one hand under a maritime influence – with both west and east coasts less than two hundred miles away – and at the same time enjoying some cooling effect from its altitude of three thousand feet. Nevertheless, come the hot months of May and June, an escape to a proper hill station makes a welcome break. So it was a particularly pleasant duty to accompany my AOC to Ootacamund in the Nilgiri Hills. (Where else but in a country like India, which borders on the magnificent Himalayas, would you find land rising above eight thousand feet being referred to merely as hills.)

With a road journey of some five hours ahead of us, the Air Commodore's bearer packed a substantial picnic hamper for our lunch, and I satisfied myself that the AOC's new driver was familiar with the route and that he had the Humber staff car in tiptop condition. For this new member of the AOC's personal staff the trip also represented something rather special, combining, as it did, an opportunity to prove his skill as a VIP driver, with the rare chance to enjoy a complete change of scene. Had he been a member of the RAF, the importance of the day would have been real enough, but for Spavarro, an Italian prisoner-of-war, it was doubly so.

The simple act of leaving Bangalore and taking the road into open country was like setting the clock back a hundred years, as the unhurried rural scene took over: here and there pairs of oxen hauling primitive wooden ploughs, prepared the ground for sowing; bullock carts meandered along in the middle of the road, their drivers long since rocked to sleep by the gentle swaying of their vehicles and oblivious to other traffic; passing under the dappled shade of a huge overhanging banyan tree, sending a score of chattering monkeys scampering to the roadside; and near a small group of roadside dwellings (little more than mud huts), several small children, naked and pot bellied, eyeing us curiously, half smiling and giving us a tentative wave as we sped past. But the clock had not stopped altogether: the bus and the lorry had arrived, and the rural scene was having to come to terms with them.

Much of the 175-mile journey was on strip roads, comprising two strips of macadam, each about eighteen inches wide, laid on the crown of the road and set apart by the width of a vehicle's track: the rest of the road surface being rolled dirt, loosely held together by a modicum of hardcore. This layout in itself introduced an element of chance for the motorist, since it obliged vehicles travelling in opposing directions to put their nearside wheels off the metalled surface as they passed one another. Lorries, in particular, seemed loath to move over until the last moment – no one could properly have accused their drivers of being 'chicken' – and the very abruptness of their swerve not infrequently sprayed the other vehicle with stones. Going through villages too could be a little nerve racking, with the pedestrians, their children and their livestock appearing to regard the centre of the street as their personal preserve as they strolled about their business.

I'm not the best of passengers myself. But with Spavarro apparently reluctant to use the horn and relying on the pedestrian or the animal taking last-minute avoiding action, the Air Commodore frequently let out a stifled groan as he thrust his foot on an imaginary brake. And it was not long before the strain of these near misses caused him to impress on Spavarro the dire consequences which might result if we knocked down a villager: informed opinion being that bystanders were quite likely to take reprisal action first and ask questions afterwards.

Spavarro, eager to please, took the AOC's words to heart. Negotiating villages from thereon, he knocked five to ten miles an hour off our speed and – like his Indian counterparts – made earlier and more frequent use of the horn. I think it was the latter action which really did the trick. The horn was distinctive enough to draw attention, and the big black car with the AOC's pennant fluttering from the bonnet parted the jay walkers almost magically. How nice it was after that, to settle back in one's seat and to leave the driving entirely to Spavarro.

The last fifteen miles of our journey found us climbing steeply along a twisting road hacked out of the hillside. As we climbed through five, six, seven thousand feet above sea level – now and then meeting thin wisps of cloud – so the hairpin bends progressively increased in frequency. By now we were within two or three miles of our destination, and our confidence in Spavarro was complete.

Just then we came upon a particularly sharp hairpin bend and, as Spavarro eased the car gently through the final stages of the turn, with the steering on full lock, a gentle tug told us that this time we had not quite made it. Poor Spavarro was beside himself with apologies. It had been the most gentle of scrapes, doing little more than removing a square inch of paint from the nearside front wing. And it had occurred within minutes of our destination. The AOC was very understanding about it, but for Spavarro it had made all the difference between being congratulated on an impeccable day's work and being commiserated with. In spite of my reassurances, he disappeared to his quarters that evening thoroughly downcast.

Arriving at Ootacamund, the climate and the lush green scenery so reminiscent of England, we were temporarily in a different world, and a most agreeable one at that. There was too the very Englishness of our hosts at Arranmere Palace and Government House to complete the transformation from life in Bangalore. Arranmere, which had belonged – or perhaps at that stage still belonged – to the Maharaja of Jodhpur, was in most respects like a large English country house, and not at all palace-like in the accepted sense. But it had at least one exotic feature: an underground passage, which led from an innocuous-looking door in the centre of the house, to the separate quarters of the Maharani about a hundred yards away. I would have been entirely innocent of its existence were it not for the fact that my chief was allocated the Maharani's rooms, and I used it to deliver his official papers.

The next day Air Marshal Sir Roderick Carr, AOC in C India, arrived at Ootacamund (or 'Ooty' as it was popularly known among the British) and was soon engaged in talks with my AOC, in what one might call this working break. For my part, I only wished they would have more working breaks if they were to be held at Ooty and similar places.

The change of surroundings seemed to bring out the best in everyone. Sir Roderick had something of a reputation for biting people's heads off, but when I bumped into him on the lawn of Government House one morning before the working day he could not have been more relaxed and friendly. By the time his Personal Staff Officer came looking for him, he was well into a dissertation on the type of grass used in the lawn, and how closely it resembled a species in his native New Zealand.

Formal dinner parties on these occasions were often the source of interesting stories, as the host or hostess sought to stimulate conversation. Our stay at Ooty was no exception. Asked whether there were any wild animals this far from the plains, our hostess recounted the complete disappearance of her dog from the lawn one night only a few months before, attributing this to a panther which, in looking back on the incident, she could recall having given its characteristic cough.

Not to be outdone, a local dignitary from the legal profession added his contribution as the meal progressed. This had to do with the scale of punishment for serious crimes – including murder itself – being subject to seasonal adjustment, to reflect the quite abnormal stress ordinary people live under at certain times of the year. He told us, for example, that in the Coimbitore area – on the plain to the south-east of the Nilgiris – there is a hot, dry wind, which blows incessantly for a month or two each year, which can lead to the build up of extreme tension in personal relationships; murder within the family being by no means a rarity.

The attentiveness of the bearer allocated to me at Government House was almost embarrassing: after all, I had not quite forgotten how to dust my feet with talcum powder after my bath. But when he offered to pack my case for the return journey whilst I had breakfast, I gave him his head. He was in any event about to receive a tip reflecting his rather special position as a Government House bearer, so I felt he might as well earn it. It was only on reaching my quarters at Bangalore that I discovered he had *forgotten* to pack my dressing gown; a handsome article in RAF colours, and much coveted by my peers. Well, how does one actively set about retrieving an article under those circumstances without creating embarrassment in high places? I could only hope that it really was an oversight, in which case the governor's ADC would devise a means of returning it. Sadly, I am still waiting.

On the return journey, being ignorant of my loss, and with my chief in a relaxed mood, I was free to enjoy the riches of the

Visit of former US President, Herbert Hoover during Mysore famine, 1946. ABOVE. The Maharaja of Mysore introducing his retinue to Lord Wavell, the Viceroy. BELOW. the Bangalore Resident introducing service officers to the Viceroy.

To Mr R. W. Jordan
Wharf House, Tone, Wellington, Somerset.

The inhabitants of Wellington, Somerset, desire to extend a very warm welcome to you on your return to Civilian Life at the termination of your period of service in His Majesty's Forces.

They also wish to record their sincere appreciation of the loyal service you have rendered to your King and Country, and to extend to you this formal expression of their gratitude together with a small gift.

On behalf of Wellington "Welcome Home" Committee.

E. L. Howard, Chairman

S. G. Glass, Secretary

A friendly welcome home message.

mountain scenery and later, on the plain, to take in the places of historical interest. At first it was the sheer magnificence of the tall blue gum trees which commanded our attention; then the spectacular view of the plain below as we snaked downwards through the hair-pins. Later it was the handsome outline of the Maharaja's palace at Mysore; and, finally, the ruins of an old fort at Seringapatam, from which the great Tippoo Sahib had stubbornly resisted the British advance upon Mysore until his final overthrow and death in the siege of 1799.

Returning to the headquarters the following day – the shadow of the missing dressing gown apart – problems which ordinarily would have niggled, if not loomed large, were easily disposed of. What a therapeutic effect the Nilgiris had had.

Parting With Old Friends

As the time drew near for my own repatriation and release, I was increasingly torn between applying to remain in the service – in the hope of securing a permanent commission – and going back to civilian life to carve out what amounted to a new career. It is no exaggeration to say that I agonised over which direction to take.

I had long since won my chief's confidence, and in my job as a PA, I enjoyed being near the hub of things. But detracting from this, in India at least, the RAF was going through an unsettled period, and one sensed it would be a while before it regained its morale and sense of purpose. And there was the generous offer from my previous employer to weigh in the balance.

In the event I took the civilian route. Mainly, perhaps, because the way ahead, while offering a healthy enough challenge, seemed a little more predictable. It also had the bonus of not delaying my repatriation to the UK and reunion with family and friends.

My last evening in Bangalore did not go at all according to plan, and I'm afraid the packing of my 'hold' trunk, which I had unwisely put off until then, suffered somewhat as a result. How else can I explain the decision to place a carton of facepowder – a present for my mother – among my clothes, where the carton subsequently exploded and Elizabeth Arden facepowder permeated the entire contents. In the early part of that evening I had strolled down to the air commodore's bungalow to take my leave, and the difficulty had been in breaking away from his hospitality as we reminisced and he had pressed me to stay for just another drink. His assurance that he would lend a hand with the packing, although kindly meant I'm sure, was hardly of practical value.

Only after two or three for the road, did I eventually get back to my room in the West End hotel to face the task. Under the circumstances, diligence, and commonsense even, seem to have deserted me. There were also the ritualistic farewells of first the *dhobi* (laundry) wallah, and then my personal bearer to be indulged before I could concentrate on the task, as they garlanded me with sweet-smelling jasmine, paid handsome compliments to my ancestors, and stood by with expectant smiles creasing their faces.

A day or so later, in mid-July 1946, I reported to the embarkation centre at Worli on the outskirts of Bombay. It was almost two and a half years to the day since I had arrived at Worli on the journey outbound from the UK. It was nice to be going home, but the captivation I had felt for India on first arriving had merely increased with time. It happens that, in recent years, I have been privileged to return many times in the course of business, and I now know that I will never tire of it, in spite of the petty frustrations. A country which can cast that sort of spell – and many foreigners share that experience – could not do so unless it had enormous character.

And what is that character? Its enormous physical expanse; the extraordinary diversity and tenacity of its people, who combine with their charming, unhurried ways a strong sense of destiny; the wide variety of its terrain and vegetation; the extremes of its climate, where an electrical storm can cover an area the size of an English county and last for hours at a time, and a sunset can outstrip even the glory and mellowness of a Turner oil painting; its exotic fruits and spices, which range in flavour from the delicate and haunting to those which will awaken the most dormant gastric juices and bring beads of perspiration to your brow; leisurely train journeys which are counted in days; its markets, full of noise, colour and tantalising aromas; the impressive relics of its past in the shape of forts, temples, mosques, palaces and public buildings; its rich endowment of wild animal, reptile, bird and insect life. All that, and I have still only scratched the surface.

As we paid off and steamed out of Bombay harbour, watching the magnificent Gateway of India gradually diminish in size, I doubt if there were any on board who did not feel a genuine affection for India, together with some anxiety for the population over the split between the Hindu and Muslim politicians which seemed to be widening by the month. Not that anyone present could have guessed at the rapid turn of events which was to occur the following year, or the scale of the consequent massacres and fleeing columns of refugees.

Our vessel was the RMS *Andes*, a trim 26,000–ton, single-funnel liner which normally plied between the UK and South America. Accommodation was no less cramped than that on my outward journey in the P & O liner. War or peace, it seemed that the common denominator for a troop transport was to move the maximum number at a time. Not that we had any complaints with that as a basic philosophy. But when some chap's got his backside in your face as he's struggling to get into his trousers first thing in the morning, and you're lying trapped in your bunk, philosophy is not your main concern.

As far as Naples, we had the company of several hundred Italian ex-prisoners-of-war, whose attention was regularly demanded on the ship's tannoy. With the lyrical sounding phrases which followed the preremptory announcement *'Attenzione, attenzione'*, and the hundreds of smiling faces looking up from the decks below, it was hard to imagine these particular men as having posed a really serious threat during the Middle East campaign. Either that, or their demeanour had softened considerably during their captivity.

The big difference evident on the return journey – the guaranteed lack of enemy action apart – was our ability to make full use of whatever speed the shop possessed, unfettered as we now were by the convoy system. This also made an accurate assessment of our speed by the average passenger a much more chancy affair, and an official sweepstake on the miles run became part of the daily routine. Conversely, losing the convoy system meant a lack of other shipping to engage our interest as we leaned over the ship's rail. But one got used to that, and it was a small price to pay for hostilities being at an end and knowing that we would get a sound night's sleep.

The peaceful nature of the homeward journey and the fact that my ties with the RAF were about to be severed, found me reflecting on my brief but (dare I say it?) enjoyable war service – above all on the good luck I had had when I most required it. It had begun with my scraping through the attestation medical, in spite of defective colour vision. It had been with me when I was commissioned regardless of the incident with the Anson undercarriage, and I had had my fair share of it during my operational tour; especially, perhaps, in finding the clearing at Shamshernagar on the day we ran out of fuel. Looking to the future, I could only hope that my guardian angel would not desert me when I put on my 'demob' suit, and that the spirited companionship of my service days would not be completely lost.

7 *A Hobby Takes Over*

A warm welcome was waiting at the family home. And the town too made its gesture by way of a small cheque and a friendly welcome-home message: a gesture it made to everyone of its returning service men and women.

Initially, the pleasure of being reunited with family and friends provided all the interest and satisfaction I could wish for. But after a few weeks, I was glad to be setting about my new career.

As I'd hoped, Fox brothers duly engaged me as assistant to their London representative, and I took up that appointment in late October. The Fox family have run a woollen business in Wellington since 1772, and in 1946 was still the town's largest employer, so I felt quite complimented to have landed the job.

My boss, George Ball, was energetic, efficient and widely respected for his knowledge of the market. Our job consisted of selling woollen piece goods to both ends of the London trade: in the West End to the woollen merchants and in the East End to the clothiers. And although very different from life in the RAF, it looked as though it would be quite challenging as more competitive conditions emerged.

Viewed in annual terms, the working routine at our Golden Square offices was punctuated by two periods of feverish activity – one at the beginning of the year, and the other towards the end – during which, each day was spent visiting two, three, sometimes four customers, to show them our range of cloths for the next season: autumn or spring. The aim being to come away with an order for despatch to the mill by that evening's post.

Those selling phases were by far the most rewarding periods of the year. True, selling in the general sense was not all that difficult, given the post-war scarcity of consumer goods. The skill and satisfaction lay in steering the customer's selection of cloths: so that, collectively, their orders would keep our production at full capacity. The customer of course knew what we were up to, and our job was to achieve our end while allowing him to gain some credit as the buyer. In its way, a process just as stimulating as poker.

But, off duty, something was lacking. And come the summer of 1948, I was ripe for a more challenging form of recreation than the occasional game of tennis, some rowing on the lake at Regent's Park, or watching the Compton brothers play cricket for Middlesex. Subconsciously perhaps, I was looking to replace the colour and comradeship which had been largely absent from my life in the two years since my demobilisation.

Taking positive steps to put this right, in August 1948 I was reappointed to a commission in the RAFVR. Apart from attending an annual camp, this was essentially a spare-time commitment, which required me to attend the occasional evening lecture and to fly Tiger Moths at weekends. As I saw it, the stimulation I would get from such a hobby could only enhance my work.

In practice something different occurred. In October 1948, I returned from my first camp an unsettled man. I had flown on thirteen out of the fourteen days, and had done so in the company of a bunch of enthusiastic young men of my own age. It had been so incredibly enjoyable, and contrasted so vividly with my daily routine, that it had been like casting off a tiresome burden. However, I was shrewd enough to know that the euphoria of the past fortnight must be allowed to settle before I seriously considered yet another change of career.

As it happened, I was denied the luxury of mulling things over. On the day of my return (a Monday), George suddenly announced that I was being given additional responsibilities and an increase in salary. He then produced a letter which Harry Fox planned to send to our Scottish clientele at mid-week, introducing me as the company's representative for that area.

What could I do but express genuine pleasure at this recognition? At the same time, I realised that I only had until the next evening to consider my position. I could hardly allow that letter to go out if there was a real possibility of my handing in my notice in the near future. The snag was that, although the RAF were taking back some demobilised aircrew, I knew it was doing so very selectively; and that applicants were having to wait for months before knowing if they had been accepted. I really was on the horns of a dilemma.

The next evening I told George of my intention to rejoin the RAF if they would have me, and followed up with a letter to Harry expressing my sincere regret at having wrecked his plans. My hobby had taken over.

Thanks mainly to the irritatingly deliberate pace of the Air Ministry's administration, it was almost six months before I reported to RAF Waterbeach on re-entry to the RAF. Meantime, George and I completed another seasonal sales campaign and I continued with my weekend flying.

My parents and friends wondered what I was up to, exchanging good long-term prospects with a reputable company for a short service commission: all the RAF was prepared to offer me at that stage. But young men are better perhaps at obeying their instincts than reflecting on the longer term. And there was more to it than the attraction of flying: great as that was. I wanted to get back to an atmosphere of camaraderie; to a life where personal initiative is encouraged; to a life where the opportunities for sport and recreation are built into the system; to a life where – at that time perhaps more than now – the trappings of the Mess would lend some style to my off-duty moments. What the RAF offered was not simply a job, it was more a way of life.

Part Two
More a Way Of Life

8   Re-enlistment

Back in Uniform
On 4th April 1949, I rejoined the RAF: retaining my wartime rank of Flight Lieutenant, but forfeiting seniority for my break in service. The formal part of the reinduction process dispensed with at RAF Waterbeach, I returned to London to discuss my future flying role.

This discussion, held at the Air Ministry, had some meat to it. Flight Lieutenant Hugh Tudor, conducting the interview, had his ideas, and I had mine. With my heavy bomber background, Hugh saw me as a natural for the Berlin Airlift, then in full swing. For my part, I'd had my fill of long sorties and straight-and-level flying, and dearly wanted to fly fighters. We compromised. I would become a flying instructor, where the sorties would be short and I would have plenty of scope to indulge in aerobatics and the like.

The outcome was not entirely as I would have wished. But, with Air Marshal Sir Basil Embry moving into the seat as C-in-C Fighter Command, Hugh made it clear that someone as *old* as twenty-six hadn't a snowball's-chance-in-hell of being allowed to make the transition to fighters. Such a rigid barrier on the grounds of age struck me as thoroughly hidebound: something I argued. But it can be counter-productive to buck the system beyond a certain point. That reached, I surrendered gracefully.

RAF Finningley ran two pilot refresher courses. The multi-engined Flight employed Oxfords and Wellingtons, and their single-engine counterpart, Harvards and Spitfires. On reporting, I was delighted to find that Hugh Tudor had arranged my course with the latter, whereby I would combine refresher flying with conversion to single-engined aircraft. Things were looking up.

The members of my course were a mixed bag. The senior students – Wing Commanders Rump and Millington – were about to take up flying appointments, following staff duties. A few junior officers and senior NCOs were switching between aircraft roles, and the remainder, like myself, were getting back into flying practice on re-entering the service.

We enjoyed good weather in our two weeks at Finningley, and made the most of it. Speaking for myself, twenty-six sorties and four and a half hours in the Link trainer in ten working days left little time for twiddling one's thumbs.

Most of my Harvard flying was concerned with practical aspects, such as the airfield circuit, instrument let-downs and forced landing practice; with little opportunity for fancy things like aerobatics. But latterly, having dealt thoroughly with the fundamentals, and before I moved on to the Spitfire, my instructor relented and we went briefly through the full gamut of manoeuvres from loops to hesitation rolls. My weekends with the RAFVR meant that I had already freed some of the rust from my aerobatic technique: but how nice it was, for the first time, to exploit the Harvard's extra power and controllability.

When I began the course, I knew the Harvard by reputation as being a lively, even skittish, aircraft in certain situations: during the landing phase for example, where if one was lax enough to touch down with any sideways drift, a damaged wing tip was a distinct possibility. As a result, I approached it with some respect. But by the end of the course I felt very much at home with it; from the front seat certainly.

As for my brief initiation in flying the Harvard from the rear seat – the position I would occupy as an instructor – it served more to illustrate the problems of doing so, rather than making me thoroughly competent in the art. In the tail-down attitude during the early stages of take-off and when holding off for a three-point landing, the view ahead was restricted, to put it at its best. From this rather limited sampling of the rear seat, it was not easy to visualise being in absolute control of the situation in every circumstance: on a dark night, for instance, with a green student at the controls. And presumably there would be limits to solving the problem by hanging one's head over the side. Naturally, I took the view that if others could do it then so could I. But it was obvious that an interesting time lay ahead.

The icing on the cake was the Spitfire, and if I had shown the Harvard respect, my attitude towards the Spitfire was initially one of awe. In the event, it handled so well that my sole concern was the speed with which the glycol temperature rose after starting up. This characteristic went a long way to explain – the nature of the role apart – why fighter pilots (those with liquid-cooled engines at least) always seemed to bolt from their dispersals the moment their engines coughed into life.

The Spitfire's true character began to reveal itself only after I'd received the green light from the runway controller. To feel the surge of power as I opened up the engine on that first take-off, and to look ahead through the blur of its four-bladed propeller as I gingerly raised the tail, gave me an enormous kick. (The idea was not to raise the tail too abruptly, otherwise the huge propeller disc would act like a gyroscope and swing the nose smartly to one side.) After gaining height, I looped and rolled for minutes on end as I thrilled to the lightness of its controls. How privileged I felt to fly that historic aircraft.

To put that period into context: not only had the Berlin Airlift (designed to bypass the Russian rail and road stranglehold) been in operation throughout the winter of 1948–49, but the deepening Cold War had also led to the creation of the North Atlantic Treaty Organisation. The treaty itself being signed on the very day I had reported to RAF Waterbeach. Of course, had there been no Cold War, I doubt if the RAF would have taken me back and, if they had, whether Hugh Tudor would have encouraged me so strongly to become a flying instructor. Put simply, the RAF was gearing up for expansion.

Scottish Interlude

To fill the gap between my refresher flying and my flying instructor's course, I was temporarily appointed as adjutant with No 666 Air Observation Post Squadron at Scone (near Perth) in Scotland: an auxiliary unit whose role was to spot for ground artillery.

Anyone sceptical of inter-service co-operation and the ability of auxiliaries and regulars to work together, should ponder the make-up of this unit. The commanding officer was a Territorial Army (TA) major, whose day-to-day stand-in and squadron flying instructor was a captain in the regular army: known as the squadron captain. The remainder of the squadron pilots were TA captains, and the aircraft maintenance and servicing crew consisted entirely of regular RAF personnel. Add to that a regular RAF officer as squadron adjutant, and you have an unlikely sounding mixture. But it worked, and worked well.

This was my first real look at Scotland. And what better way than from one of 666 Squadron's Austers? David Burroughs, the squadron captain, gave me a familiarisation flight and, having gone solo, I was able to fly two or three times a week – sometimes on squadron business to RAF Turnhouse at Edinburgh, and on other occasions to air test an aircraft, or just to stay in flying practice.

Flying in the vicinity of Scone airfield was fine for the latter purpose, but with the need to keep an eagle eye open for other aircraft, it provided little opportunity to sit back and admire the beauty and grandeur of the Scottish countryside.

That was where the flights to and from Turnhouse had the edge. Leaving Scone, the first of the en route features to make an impact were the Ochil Hills, which rose to twelve hundred feet on track and over seventeen hundred feet just to the east. Since the whole leg was only thirty-five miles or thereabouts, it was hardly worth climbing to normal cruising height, so provided the visibility was good, I usually indulged in a bit of harmless low flying as I cleared these handsome, craggy obstacles with only a few hundred feet to spare.

Clear of the Ochils, and given some breaks in the cloud, the next feature to reveal itself was the bold shape of Loch Leven, with the outlines of the Burleigh and Lock Leven castles to add a touch of nobility to the scene.

One's arrival at the southern end of the leg was preceded by the gradual unfolding of the Firth of Forth, studded with craft of various kind; and the emergence of the unmistakable outline of the Forth Bridge – a mere five miles from Turnhouse airfield, and conveniently pointing almost directly at it. With the River Tay and the city of Perth to guide me towards Scone airfield on the return leg, navigation on those Edinburgh visits was hardly demanding.

On one occasion, I accompanied David Burroughs on what he termed a 'Field Reconnaissance', and which involved a more deliberate exploration of the countryside. This took us over a wonderfully untamed mixture of mountains, forests and lochs in the area to the south west of Perth, where one stunning view followed another as we weaved our way through the mountain passes and dipped down to skim along a hundred feet or so above the lochs. With David at the controls practising the art of low flying, I was free to drink in the scenery and to keep an eye on the map. But David knew the countryside, and my only influence on our route was to suggest a diversion to Loch Lomond and its 'Bonny, Bonny Banks', so that I might sample what lay behind the words of that lilting song.

We were a lodger unit at Scone airfield, the main user being Airwork Services, who operated an elementary flying school and also staffed the airfield services. At that time, Airwork was equipped with Tiger Moths, and a bonus of our good relationships was being able to borrow a Tiger Moth to keep my hand in at aerobatics. Not only was the Auster not cleared for aerobatics, but if you as much as subjected it to two or three sustained steep turns, you were as likely as not to return with a buckled engine cowling: presumably due to overstressing. And to make a habit of this was no way to endear oneself to the ground crew.

Once a week David and I had lunch with his brigadier in the George hotel, at Perth, whose elegant dining room overlooks the Tay. These occasions served mainly to keep Scottish Army Command briefed as to what we were up to. But they also kept our finger on the pulse in the reverse sense. And what more pleasant way of doing so than over a nice piece of salmon and a glass of Chablis.

Yes, Scone was a pleasant backwater: what with the magnificent countryside; the engaging forthrightness of the local people; comfortable living quarters with a squash court alongside; the chance to stay in flying practice; the undemanding working pace and so on – but a backwater none the less. When I received my posting to the Central Flying School for my flying instructor's course, it felt good to be moving on. I reported to RAF Little Rissington on 20th July.

ABOVE. Pilot refresher course, 1949. A Spitfire XV1, the 'Icing on the cake'. **CFS 'Type-Flying' experience.** LEFT. Meteor T7. *(next page):* ABOVE. Mosquito, BELOW. Lancaster. *(Photo's courtesy of Avro International/Ian Lowe)*

9 *Flying Instructor*

The Course

We were welcomed on arrival at the Central Flying School (CFS) with a drinks party – a civilised way of breaking the ice with our instructors. I had joined the one hundred and twelfth flying instructor's course: over fifty strong, with more than a fifth of our number coming from the air forces of India, Pakistan and Egypt.

CFS was established in 1912: in aviation terms, almost at the beginning of time. Its original task was to produce military pilots, but since 1920 it has been responsible for training the RAF's flying instructors. And with the growth of its reputation in the instructor training role, for many years it has also performed that function for a number of foreign air forces.

At units with such an illustrious background, perhaps a certain self-consciousness and swank is inevitable. CFS had something of both. As the course progressed, however, I found that this pretentiousness was more than offset by the individual professionalism of the staff. And in their offices – divorced from those yellowing photographs which lined the corridors of the Mess – the attitude of those in charge was vigorous and forward looking.

One of the pillars of RAF doctrine is that if the pilot is to fly his aircraft to its limits – so equipping himself to get the best out of it in combat – he must know what makes it tick. Similar importance being attached to a range of other matters which can influence the outcome of his sortie: his physiological limitations, the weather, navigation techniques, and so on.

Putting this philosophy into practice means that every trainee pilot divides his time, more or less equally, between classroom lectures and flying. As budding flying instructors we followed the same routine, and at the end of the course we sat examinations on the full range of classroom subjects, as well as having our flying and instructional ability rigorously tested.

Immediately after our rather posh reception at Little Rissington, we were whisked off to RAF Brize Norton – about a dozen miles away as the crow flies – where we spent the first two months of the course cutting our teeth in the art of talking and flying at the same time. Talking not just any old blarney – I suspect many of us were not bad at that already – but synchronising an instructive commentary with a convincing demonstration.

The instructive commentary, or 'patter' as it was known, was not left to individual inventiveness or whim. The points to be put across were enshrined in an Instructor's Handbook, which dealt with the gamut of air exercises required to take the student step-by-step from his familiarisation flight to the presentation of his Wings: ranging over such diverse control techniques as the subtle handling required to maintain straight and level flight, to the full-blooded application of the controls when recovering from a spin.

The instructor's handbook also set out the contents of the pre-flight briefing for each exercise. But although it would be available to us as Qualified Flying Instructors (QFIs) as an *aide-mémoire* for discreet consultation, we were strongly urged to commit the briefings and air exercises to memory as we went along, and to avoid the handbook becoming a crutch.

Sound advice I'm sure, but I suspect that a few of our more enterprising colleagues smuggled the handbook into the air to help them through the periodic tests. In my own case, I left it behind not so much from high principle but because its mere presence might serve only to disrupt such natural flow as I had acquired.

My flying instructor was Flight Lieutenant 'Mack' Mackilligin: a big, hearty, pipe-smoking man, who loved a party. Four of us shared Mack's expert attentions. Off duty, Mack liked nothing better than piling us into his car and heading off into the Oxfordshire countryside to down a few pints. This might not have contributed to clear thinking the following morning, but it did a lot for staff/student relations.

The pattern for each teaching exercise was that Mack first gave us the pre-flight briefing, then took each of us through the exercise in the air. We students then paired off and took it in turn to give the pre-flight briefing, and to 'patter' the exercise in the air; swopping seats between the flying sorties so that each of us played the instructor and student roles in turn.

When Flying Officer 'Dougie' Palmer and I first tackled these air exercises, the one playing the role of the instructor dried up or lapsed into waffle every few minutes. But by constant repetition, these embarrassing pauses and verbal fumbles gradually disappeared, and even *we* began to see ourselves as instructors. And as for job satisfaction, I doubt if those of with even the smallest spark of theatre did not enjoy a degree of this when the occasional exercise went off as per book.

Given the absorbing nature of these sorties, each of which lasted an hour or more, it would have been all too easy to lose track of our position. How grateful we were in these circumstances that Blenheim Palace was in the Brize Norton local flying area. As long as we kept its handsome symmetry in view, and that of its great avenue of trees, we knew we were within a few minutes of the airfield: a useful aid for getting the most out of our sorties in deteriorating weather.

In late September we moved to Little Rissington. By which time my hours in the back seat of the Harvard, in daylight at least, were well into double figures, and I had also had an hour at night, practising take-offs and landings from the back seat. All of which had been accomplished without as much as scratching one of His Majesty's aircraft. My initial scepticism about the poor forward view from the rear cockpit had softened considerably by this time. It really is extraordinary what one can get used to.

I had last been stationed at RAF Little Rissington in 1943, on return from my Wings training in Canada, and the local villages and their inhabitants seemed to have changed little meantime. One of my off-duty pleasures in the weeks which followed our move was to renew some of my earlier connections. How reassuring it was to find that the Old New Inn in the village of Bourton-on-the-Water and the Bull at Burford still made one feel personally welcome, and that the locals had not traded their charming, unhurried ways or the soft burr of their Gloucestershire accents for more dynamic and upmarket equivalents. If ever we felt staleness creeping on, an evening at one of the local pubs was a great reviver.

The airfield at Little Rissington, standing at 750 feet above sea level, has the distinction of being the highest in the British Isles: a distinction not without its disadvantages in certain weather. A fact which was borne in on us from time to time, and which frequently kept the supervisory staff on their toes.

Perhaps the most critical situation facing the staff was when the country generally had a layer of early morning fog, with Rissington unaffected because of its altitude. Under these circumstances, the nearest diversion airfields were probably Valley in Anglesey or St Mawgan in Cornwall. The question was, although Rissington was basking in brilliant sunshine, should they risk launching their aircraft, knowing that before there was a general clearance of the fog, the visibility at Rissington itself might temporarily deteriorate as the fog 'boiled' upwards from the surrounding valleys?

On other occasions, Rissington was the odd man out in the reverse sense; retaining bad weather long after it had lifted from other airfields in the region. For one reason and another, life was seldom dull during those four months at Little Rissington.

In the middle of November, I had an experience which truly lifted life from the ordinary – my first jet flight. Opening up the Meteor's twin engines on take-off, one was thrust into the air at an incredible rate. And overshooting from a roller landing – with the aircraft already rolling at over a hundred miles an hour – the acceleration was even more breathtaking, as one's seat was shoved hard up against one's back.

In fact, the jet aircraft has such a smooth aerodynamic shape, that one of the problems facing the aircraft designer was to devise an effective means of *losing* speed: mere throttling back – in the absence of external propellers – having little effect at high speed. So he came up with the airbrake: hydraulically operated flaps designed to break up the airflow. And how very effective these were when operated at high speed on the Meteor; the aircraft decelerating so rapidly that our shoulders were thrust forward hard against the safety harness.

But even the Meteor was limited in how fast it could go. And part of my familiarisation flight was to demonstrate those limitations. As we approached about eight-tenths of the speed of sound, and shock waves began to attach themselves to the aircraft, it was like leaving smooth macadam and running on to a cobbled road. The buffeting was quite severe and accompanied by a marked change in trim. For me, the experience was entirely new: in technical jargon, we were experiencing 'compressibility'. To recover, we merely popped out the airbrakes and throttled back: almost immediately, we were back on smooth macadam.

Long before that first jet flight was over, I knew I had tasted flying of an entirely new character. However, everything has its price, and it seemed to me that the jet's electrifying speed and acceleration were more than matched by its thirst. You could practically see the needles of the fuel contents gauges fall, even as you glanced at them.

The Meteor in question was the Mk 7 which, like the Harvard, had two seats in tandem. There the similarity ended. The Meteor had a tricycle undercarriage, which did away with the need to swing the nose from side to side when taxiing, and gave even the man in the rear seat a good view ahead. And with no propellers in the conventional sense, the Meteor had no in-built tendency to swing on take-off. It was in fact remarkably uncomplicated to fly, and although my first jet flight was under instruction, I probably handled the controls for a good fifty per cent of the time. Two further sorties, and I was sent solo: only in the circuit it's true, but enough to provide some sense of achievement and to leave me wanting more.

These jet experience flights were part of the CFS policy of broadening our experience as fledgeling instructors, so that when we took up our duties as *ab initio* instructors, we would be that bit more credible in the eyes of our students.

Other aircraft we flew in this context were the Lancaster and the Mosquito. The Lancaster, being fundamentally similar to the

Liberator in having four engines, did not provide any great revelations for me personally, except to impress me with its sprightly take-off and easy manoeuvrability. The Mosquito, on the other hand, had a much more challenging temperament, and had to be handled accordingly: it was a particular source of satisfaction to fly it solo for the first time.

Towards the end of our course, by way of broadening our perspective, we visited the Bristol Aircraft Company at Filton. Much of our attention being devoted to the Brabazon airliner: a huge aircraft by the standards of the day, but (due to lack of an engine with the right balance of power and fuel economy) already being spoken of as a white elephant. However, looking back on its sad demise, and to the fate of the Comet when it was first introduced as a civil airliner, one can hardly accuse the British aircraft industry of being unenterprising.

On 13th January 1950, I ruled off my log book and went on embarkation leave. Meantime I had passed my ground school examinations and qualified as a flying instructor with a B1 category – the higher of the two categories obtainable. I had been posted to No 4 Flying Training School (FTS) in Southern Rhodesia; the prospect of further overseas travel pleasing me enormously. We had also been formally dined out, and sent on our way with all the good wishes of the oldest unit in the RAF.

On that final evening – the silver trophies adorning the tables glowing softly in the mellow light of candelabra; a medley of tunes billowing down from the minstrels' gallery; the courses arriving and being cleared with quiet precision; and an excellent meal being rounded off with witty speeches over the port – it was time, once more, to take stock of CFS. My first impressions, six months earlier, of a unit perhaps too conscious of its past, had mellowed. Since then, I had seen for myself that the way it cherished its heritage was only one side of the coin, and that it was equally determined to stay abreast of flying techniques in the best possible way: that is, by getting its hands on the latest hardware. I concluded that any unit which could tread that delicate path between tradition and progress with such obvious success had little wrong with it. And the fact that it could run such an impeccable and hugely enjoyable dining-in night merely underlined its professionalism.

The RMMV *Athlone Castle*

On 16th February 1950 I set out for Cape Town: one of a small RAF contingent on the first leg of its journey to Southern Rhodesia, now Zimbabwe.

In those blissful days before the jet airliner, the majority of service-men and their families went overseas by ship. No question of being hurtled through time zones with brutal efficiency and arriving at their destination dazed and bewildered. No, we did things at a civilised pace. In this case, what's more, we were tacked on to the passenger list of a luxury liner, the RMMV *Athlone Castle*: one of a fleet of beautifully appointed Union Castle vessels which plied between Southampton and South Africa.

Sharing my cabin were Flight Lieutenants Bill Tait and Ken Thompson. Bill, an ex-RAF apprentice, was a wily forty-year-old, with a deeply lined, leathery face. Although Bill and I had been at CFS together, I knew surprisingly little about him since he'd lived out of Mess. As for Ken, he was a year or so older than myself, and I'd only met him the night before we sailed: he had a certain aloofness, or perhaps it was a natural reserve – either way, he was clearly going to take a little getting to know. But at least we all had one thing in common – each of us was bound for 4 FTS: Bill and I as flying instructors, and Ken as a lecturer in the principles of flight. With fourteen days at sea and a thousand-mile rail journey ahead of us, chances were we would be well acquainted before we reached our destination.

The RAF contingent included a senior NCO's wife travelling without her husband, but accompanied by her thirteen children. As OC Troops, Ken Thompson initially looked a little askance at the size of this family and the problems which might fall into his lap en route. But, except for a little help from his imprest account as the voyage progressed, I believe the demands of Mrs Taggart and her spick and span little troop were practical nil.

Sea travel was by no means a new experience for the members of my cabin, but none of us had previously travelled first class under full peacetime conditions. How nice it was to have reasonable cabin space for a change, and all the amenities of a cruise liner too.

Our passage through the Bay of Biscay was not unduly rough as I recall, but it was not until we had cleared that area that the more timorous passengers were appearing regularly in the dining saloon. And it took the warmer air south of Gibraltar to set the ship's social routine properly alight.

The turning point came when the ship's orchestra appeared on deck for its evening session, and the boat deck became the dance floor. Led by a rather seedy-looking violinist, whom we unkindly dubbed 'Scraper Sid', they were certainly not the most scintillating of dance bands: had they been, presumably the demand for their services would have kept them nearer home. But at least they gave all they had and, after a few nights, their very doggedness had won us over.

Sid's deferential manner and his limp smile hinted at years of scraping away in the service of others. And although his blotchy complexion suggested regular access to strong drink; in fairness, I think that to have done his job every night in complete sobriety would have been asking rather a lot.

So it was that after an hour or two of solid application, Sid would murmur a few words of apology and slip away with the rest of the band for a thirty-minute break. These silent interludes did not remain as such for more than a night or two, however: with Sid's permission, and with the growing approval of the passengers, Ken Thompson took over the pianist's stool on most evenings. What Ken's full repertoire comprised I can't remember, but you could put your shirt on him playing *I'd like to get you on a slow boat to China*.

The lyrics of that particular tune seemed especially suited to Ken's frame of mind. Within forty-eight hours of sailing, he had homed in on a pretty girl named Moyca, who was travelling with her parents on a visit to her brother in Natal.

Moyca had a bubbly, outgoing personality and it was apparent that she had also attracted the attention of one of the members of the ship's crew – a presentable young man with bright red hair. For some days Bill Tait and I, and doubtless others, wondered how this contest would end.

By the time we anchored briefly in Madeira, the red-haired lieutenant was losing ground fast. But the question remained, with over a week to go before we disembarked at Cape Town, would Ken's flourishing romance run out of steam?

Bill Tait meantime was befriended by two attractive ladies from Yorkshire. Elsie and Reanie (also fortyish I would guess) seemed quite happy to share an escort. And Bill, perhaps seeing safety in numbers, clearly enjoyed their company: and for that matter, their champagne as well. In fact they seemed to drink little else, and it was not long before he was giving Ken and me a thoroughly convincing rendition of *Champagne Charlie* on his return to the cabin.

After Madeira, the gracious end-of-voyage ball apart, the major social event was the ceremony of 'Crossing the Line': an elaborate affair staged by the ship's crew with some assistance from a few specially briefed passengers, of which I was one. Rather to my surprise, our briefing laid great stress on the need to avoid spontaneous horseplay, which it was said might easily end in a brawl. In the event, we managed to keep the lid on things, but it was rattling a bit at one stage, by which time it was clear why we had been briefed with such care. With the little indignities the ceremony called for, and with such an expectant audience, there was every temptation to improvise.

Disembarking at Cape Town, most of us spent the day walking the streets, finding our land legs and getting the feel of the place. Not so Ken and Moyca, who were discussing their future somewhere on the slopes of Table Mountain.

That evening we began a two-and-a-half day rail journey to Bulawayo; Ken, Bill and I again sharing accommodation. As Ken boarded the train and slumped down in his seat, he was patently suffering the pangs of separation: a condition which appeared to affect him for most of the journey. But his romance was by no means a lost cause.

Moyca was due to spend an extended holiday on her brother's farm at Greytown in Natal, a mere thousand miles from Bulawayo by road – not an unbridgeable gap as distance is regarded in those parts. En route to Bulawayo, Ken was already totting up his capital assets and talking car prices.

In terms of distance, our rail journey was reminiscent of those I had made in India, and some of the scrubland was not unlike that of the south Indian state of Mysore, with its granite outcrops and sparse vegetation. The efficiency of the South African railway system too was reminiscent of the high standards maintained by the Indian railways. As to the difference, perhaps the most striking contrast with my memories of India was the neat and well kept appearance of the African native settlements, and a general absence of the begging hand.

4 FTS

We found RAF Heany in scrubland about sixteen miles out of Bulawayo. Far enough in the sticks to impose a measure of social independence; and among the better off, to encourage car ownership.

The station commander, Group Captain Andrew Geddes – a Scot, and proud of it – personally addressed each batch of new arrivals; devoting something like an hour to describing Rhodesia and its people, and to highlighting those adjustments we would need to make if we were to get the best out of both. At one stage he gave a graphic account of how the bilharzia bug can enter the bloodstream through simple abrasions and attack one's liver: the maxim being – never bathe in stagnant water. Although rather horrified to learn of such dangers, I was more than glad to be warned of them.

Andrew began his service career in the Royal Artillery and had transferred to the RAF as recently as 1945. Before his transfer, he had been seconded to the RAF in the Army Co-operation role on no fewer than three occasions, so that by the time the permanent switch came he was no stranger to 'light blue' habits. But with his clipped moustache, his brisk manner and his penchant for horse riding, he was certainly making no apologies for his pedigree. There was also a tough streak in Andrew's make-up: something which no doubt played an important part in the work he did in aiding the resistance groups after the fall of France, and for which he was awarded the DSO. The net result was a tight ship.

The 4 FTS crest, with its palm trees and pyramids, told immediately of the unit's longevity and of its links with Egypt in earlier times. And later, as we delved into its history, we learned how it had turned its hand to operations during the 1941 uprising in Iraq. A unit's history may be no guarantee of its present efficiency, but there is no denying that the fine collections of silver which units such as 4 FTS amass along the way add a certain glamour to dining-in nights. Happily, 4 FTS's working standards at Heany seemed fully in keeping with its illustrious past.

I began my instructional duties with the Tiger Moth squadron, where we prepared the newly arrived cadets for their first solo flight. As I helped others towards that thrilling moment – one I had experienced some eight years earlier – the wheel had come full circle. As a job, I cannot remember one more satisfying.

Those cadets who went solo did so generally without incident, other than to be 'on cloud nine' as they rejoined their friends afterwards. And only now and then did we suffer palpitations as we watched a cadet on his first solo seemingly forget all he knew about landing the aircraft. After one such man had overshot for the fourth time from a perfectly sound approach, I began to wonder just *what* was holding him back. However, I'm glad to say he eventually got down in one piece. He too showed some satisfaction as he climbed down from the cockpit: as much from relief as anything, I fancy.

The Rhodesian seasons differ from those of the UK. The winter months, from April to November, are the most equable: bringing little rain and comparatively cool temperatures, while the summer months, from December to March, are hot and wet: producing turbulent flying conditions, and thunderstorms of real severity.

So, for the eight months of the Rhodesian winter, our working hours were similar to those at home. The big difference was that, in Rhodesia, we could get on with our work with little hindrance from the weather. Almost the only problem was an occasional band of low cloud – known locally as 'Gooty' – which usually formed in the small hours. When this happened during night flying, it was often a matter of nice judgement just when to recall the aircraft before the visibility became dangerously low; sometimes the Gooty would threaten the circuit traffic for an hour or more before suddenly closing in. Local knowledge was useful in these circumstances, and at least one officer in charge of night flying – watching the situation from the control-tower balcony – was known to order the aircraft to land immediately the handrail had acquired a certain dampness.

In summer, to minimise the effects of the heat and rain, we began our working day at six o'clock; finishing at one o'clock as we approached the hottest part of the day. To get the flying programme off to a prompt start, a proportion of the staff and students went to work on a cup of tea, breaking off at around eight for a cooked breakfast, by which time the rest of the staff and students had arrived to maintain the momentum.

In spite of the early start, there was the keenest rivalry to see which Flight could be the first into the air: my Tiger Moth Flight Commander, Flight Lieutenant Vere Potts, regularly lending the ground crew a hand to swing the propellers.

Flying on that first shift, with the sun only a few degrees above the horizon, conditions were incredibly smooth. Not only that, somehow the whole scene was more tranquil – the distant hills of the Mulungwane Range were shades bluer than they would be by mid-morning; small overnight pockets of mist still persisted; and the smoke from the townships and small mining communities round about rose vertically, as yet undisturbed by the sun's heat. Yes, there *were* compensations for those hurried ablutions and that early morning dash to the airfield. No doubt our students too were glad to swop the turbulence of midday for the early morning smoothness, especially if they were 'under the hood', flying on instruments.

In 1951, Heany's Tiger Moths were replaced by de Havilland Chipmunks: all-metal monoplanes with enclosed cockpits. To lose the Tiger Moth was a bit like saying goodbye to a faithful vintage car: you knew it had to happen eventually, but when it did, it removed an element of romance. Gone was the wind on your face;

and gone also the gentle whine from the rigging as you throttled back for a glide approach. But by then I had transferred to the Harvard squadron, in which the students completed their training to Wings' standard, so the wrench for me personally was at one remove.

Later, I was to experience a more practical problem thrown up by the Tiger Moth's demise. An *ab initio* cadet was handed over to me after only one flight in the Chipmunk, because, at six feet seven inches, he was too tall to fit into the cockpit. My mandate was to train him from scratch on the Harvard. My logbook shows that Acting Pilot Officer Dean required less than fifteen hours to solo: an extremely creditable achievement on his part, given that he was one of one blazing that particular trail. Had the Tiger Moth still been in service, of course, his head and shoulders could have happily stuck out into the slipstream.

There is no doubt that the outbreak of the Korean war in 1950 boosted the call for more trained pilots. The effect on 4 FTS was that we flew as many hours as we could sensibly cram into the day. And although for eight months of the year the weather was strongly in our favour, the flying target was adjusted to take account of that favourable weather factor; so the pressure never really eased up the year round.

This accent on quantity also caused our masters at the Group Headquarter to review the permissible wastage rate, and in some instances where we had recommended that a cadet be suspended, to ask for extra instruction to be given so that the 'borderline' case might be squeezed through. Although not blind to the underlying politics, we did wonder at times if we were not passing on to the Jet Advanced Flying Schools a percentage of pilots who might come unstuck on the more demanding Meteor.

Our working tempo was essentially a brisk one. Each instructor had up to four students, and flew with all of them in the course of two working days: usually flying three to four one-hour sorties a day. But when the students entered the night-flying phase towards the end of their course, we sometimes flew as many as six or seven sorties in the working day, spread over the day and night shifts.

There was, of course, no question of the students themselves being asked to match that sortie rate. And there were rules too for avoiding dangerous levels of fatigue in the instructors, whereby, if we landed after midnight, we did not report for flying until say ten o'clock the following morning. And, in winter, if the final landing was after one o'clock, we did not report for duty until, after, lunch. Typically, QFIs flew thirty to forty hours a month, and exceptionally fifty to sixty hours.

As mentors, the duties of the living-in officers in particular went beyond the flying programme and into the realms of etiquette. The senior students shared our Mess and, as the weeks went by, they were expected to absorb our standards of behaviour: a fact we members of staff tried hard to bear in mind. Largely because of this requirement, we dressed formally for dinner on four nights of the week: wearing a stiff-fronted shirt, a wing collar and a black bow tie.

These dress regulations may sound pretty stifling, and at first there were moments when we cursed roundly as we struggled against the clock to produce a decent knot in our bow ties. But once we had got into the swing of it, this nightly dressing up became just another part of the routine.

On formal dining-in nights, held once a month, the ratchet was taken up a notch or two, and we wore our miniature medals and decorations; foregathered for sherry; walked into the dining room to the beat of the station band; and did not smoke until after the loyal toast – and then, only with the Mess president's spoken permission.

As well as its instructional task, Heany also played an active part in Rhodesia's air-search organisation: a system for locating missing aircraft. It was at the end of one of these search missions that I experienced one of the most freakish incidents in all my years of flying.

Twenty minutes after taking off from Salisbury's Belvedere airport on the return flight to Heany, I happened to look back over my shoulder and was astonished to see my peaked cap stuck fast half way up the Harvard's fin: held in place solely by the airflow. The only possible explanation was that it had been sucked out through the open canopy on take-off.

Suddenly I had a real challenge on my hands – to land at Heany without losing that hat. With the best part of two hundred miles to go, and the landing to make, that seemed a tall order. At first I was almost afraid to shift position in my seat, I was so concerned not to dislodge the wretched thing. However, by avoiding other than the gentlest of manoeuvres, and by making a powered descent and landing, I was able to hold it in place until it flopped gently to the ground at the end of the landing run. Not losing that cap was an extraordinary piece of luck by any standard; given that the odds of it wrapping itself obligingly around the fin in the first place must have been something like a thousand to one. How kind the gods can be!

During 1951, Andrew Geddes was succeeded by Group Captain 'Bob' Constable-Roberts; a warm-hearted man who very much wanted the affection of those serving under him. 'Happy Heany' was his catch phrase, and he worked with enthusiasm to make it a reality. Neither were his efforts confined to superficial matters. In his time, the airmen's gymnastic and sports facilities underwent a major structural improvement, with a good deal of the work being done on a self-help basis. The latter way of getting things done by no means met with the approval of the Ministry of Public Buildings and Works (MPBW), who felt their role was being usurped. But once a gymnasium has been put up in the name of the station commander, it is a brave official who insists on it being taken down.

Bob and his wife 'Vella' were also extremely sociable, and gave some memorable parties. One I recall was a fancy dress occasion, where Bob appeared as Nero, with a bedsheet for a toga. The party went on until the small hours and was rated a huge success. But there was one minor hiccup. Shortly after midnight, one senior officer was obliged to retire with what his wife described as 'a touch of malaria'. Her explanation for his sudden departure might have been entirely accurate, but in the weeks that followed, I'm afraid that tag was often mischievously applied.

It was not often that Heany put upwards of twenty Harvards into the air in a single formation, and rarer still for the crews to find it an occasion for mirth: well, mirth for most of the crews, let's say.

But such was the case when we were rehearsing a formation fly-past in which the object was to spell out the letters R A F.

To help us achieve the best positioning, two of our colleagues flew in another Harvard acting as a whipper-in. Until their R/T became unserviceable they came up with instructions like 'Close up, red three' or 'Blue nine, you're too far forward', and so on. No humour there, but what our two colleagues did not realise was that after they stopped receiving any R/T response from the formation – and assumed that their external communications were unserviceable – the whole of their subsequent intercom discussion was broadcast to the rest of us. It was at that stage that we learnt what they really thought of our efforts, including those of their squadron commander who was having a little difficulty with his station keeping.

By the time we landed, the pair of them were already in the bar, having finally abandoned their attempts to contact us. When we confronted them with a few quotations from their *private* discussion, and remarked that their squadron commander would be along at any moment, they all but choked on their beer.

We also had our sad moments. One of which was a reminder that no aircraft is a toy – not even the primitive Tiger Moth. As I landed from a Harvard sortie, my student was hastily bundled out and replaced by a medical orderly: my instructions being to get him to the grass strip at the Mielbo bombing range with all speed. Within ten minutes my passenger was running the last few yards, first-aid bag in hand, towards the crumpled wreck of a Tiger Moth.

By then, the Range Safety Officer had the occupants clear of the wreckage and lying in the shade of a makeshift canopy: made from broken struts and one of the parachutes. Both men were badly injured; one groaning quietly, while the other – the one with the more serious injuries – talked almost incessantly. The orderly injected painkillers, and slowly the noisy one's protestations grew fewer. But from time to time he burst forth again: his theme – he wasn't going to die, he was too tough for that. Given his condition, the poor fellow's defiance was both admirable and pathetic. But there was no way he was going to give in easily, and he survived the long road journey to the hospital; only to die shortly afterwards. The

CFS 112 Course and the Brabazon.

ABOVE. 4 FTS Tiger Moths (RAF Heany below). RIGHT. 4 FTS Harvard IIA entering a loop *(Photo courtesy of Gp. Capt. Jack Holt).*

other one – a fitter being flown in to carry out some maintenance – did survive, but endured months of hospital treatment. The cause of the crash – the pilot's failure to recover from an aerobatic manoeuvre at low altitude.

To maintain the purity of instruction, our performance as instructors was monitored at intervals. Occasional flights with one's squadron commander apart, closest to hand in this role was the 4 FTS Standardisation Flight, which could pick on individual instructors at any time to see that they were adhering to the standard briefings, patter and demonstration techniques. But the big event of the year in this business of monitoring was the visit of the CFS Examining Team (known unofficially as 'The Trappers'), ably led in my time at Heany by Wing Commander John Barraclough, who combined ability with oceans of charm.

The team had two functions: to test the instructors, and to sample the product. In the case of instructors, either to check for standardisation of technique or to see if the individual was worthy of upgrading to a higher instructional category. The full range of categories being: B2 – Below Average; B1 – Average; A2 – Above Average; and A1 – Exceptional.

Securing one of the higher categories did not mean more pay or promotion in the direct sense. But, apart from the personal sense of achievement, such upgrading was noted on the individual's record and was generally regarded as a useful bonus in enhancing his career. So, quite apart from the reward of seeing one's personal students make progress in the everyday sense, there was no shortage of stick and carrot to maintain the instructor's interest. And as far as I personally was concerned, this mixture seemed to be quite effective – in my second and third years at Heany, I was upgraded in turn to the A2 and A1 categories: pause to blush!

This might be the right moment to get another personal achievement out of the way (well, this is an autobiography). Whilst at Heany, I was also appointed to a permanent commission. No thanks to the medical authorities, however. Once again they demurred over my defective colour vision. It was months after my colleagues knew of their fate that I was advised of an 'Air Council' ruling in my favour: the letter making it clear that this was in spite of my being 'below the medical standard required'. Meantime, presumably there had been something of a tussle between the medical and general duties branches. While recognising that the medical branch had a job to do, how glad I was that the last word lay with officers of my own branch, who took a more pragmatic view. Clearing that hurdle lightened my step for days.

A Crack on the Head

Heany, like most of Zimbabwe, is on a plateau about four and a half thousand feet above sea level, with the result that for a week or so after our arrival, any serious exertion quickly had us out of breath. But once acclimatised, the sheer predictability of the dry season found us eager for outdoor exercise.

With the rugby season about to start, I was ushered into the station team: packing left in the second row of the scrum. Later in the season, I found myself selected to play for the RAF Rhodesian Air Training Group (RATG). This accomplishment may sound rather grand, but with only two stations and the headquarters staff to draw on, the RATG selectors were hardly swamped with talent.

The pace of rugby in Rhodesia is pretty killing, and you need to be thoroughly motivated if you are to perform at all well. The dry ball not only encourages fast, end-to-end play by the three quarters, but also makes those immense kicks for touch quite commonplace, both of which frequently had us forwards running from one end of the ground to the other before we could get on with the real work of scrummaging or contesting the line-out. It was in Rhodesia that I first experienced tunnel vision. Forty minutes each way was the norm, and in the last fifteen minutes of a hard-fought match one's peripheral sight was practically non-existent.

Unfortunately for me, my rugby-playing days were cut short only five months after arriving in Rhodesia. The crunch – all too literally – occurred at Livingstone in a game between RATG and the locals, when I took a full-blooded tackle into touch. Only recently, someone confronted me with the words, 'My God, Reg Jordan, isn't it? The last time I saw you, you were lying unconscious on the touchline at Livingstone – we were quite worried about you.' That injury, which initially put me into the local hospital for a week, eventually cost me two months out of the cockpit.

Having talked my way out of the Livingstone hospital and arranged for Heany to collect me in a Harvard, I quickly realised that I had by no means kissed goodbye to the after-effects of concussion. On the flight to Heany, it was only too apparent that the shaking up my brain had taken had left me unusually sensitive to the din of the engine. Within a day or so, I was plagued with headaches, and quite soon I was on my way to a rest home at Cape Town.

'Raiders' Rest', as the home was quaintly named, attracted an odd mixture of convalescents: more than one on a short fuse, I noticed. Apart from a fellow officer from Heany, there were Naval ratings, a social worker, and – one of the biggest handfuls for the staff – an alcoholic journalist. When sober, the latter was a charming and intelligent man, who fairly whistled through the *Cape Times* crossword after breakfast. But once on the booze he could be thoroughly bloody minded. His evening meanderings were a constant worry to those in charge, and as lights-out approached, I was co-opted almost as a matter of routine to winkle him out from one of the local bars.

By the time I returned to Heany, Ken Thompson was about to set off for Natal and his wedding to Moyca. And as I was to be his best man, he welcomed me with some relief. Since arriving in Rhodesia he had bought himself a brand new Morris Minor, all of 850cc, with which he was about to tackle the thousand-mile journey: quite a bit of it over dirt roads and their alarming corrugations.

A few days after Ken's departure, I followed him by air. He and Moyca being married in the C of E church at Greytown in late September 1950. Once I had handed over the ring, my only moment of anxiety occurred as they left their reception some hours later. At which point they nearly finished up, car and all, in their host's outdoor swimming pool. But nearly is nearly, and after we had recovered our breath, many of us agreed that Ken's last second

swerve had lent a certain panache to their send-off; and that it
confirmed his pilot's touch.

The Rhodesian Air Training Group

RAF Heany was part of the Rhodesian Air Training Group (RATG):
an entirely RAF organisation, with headquarters at Kumalo airfield
on the outskirts of Bulawayo.

The only other RATG flying station, at that time, was at
Thornhill, near Gwelo. Those of my readers familiar with the
Second World War Rhodesian element of the Commonwealth Air
Training scheme will no doubt recall a more lavish training structure.
But that is not part of my story.

When I arrived at Heany, in March of 1950, Thornhill housed a
navigator training school. But later in my tour of duty – possibly
the following year – Thornhill's role was changed to that of pilot
training, this being accompanied by some stripping out of Heany
talent to help with the change. Squadron Leader Jackie Holmes,
who had commanded the Harvard squadron at Heany, was posted to
Thornhill to command the flying wing, and took with him as two
of his Flight Commanders, Flight Lieutenants David Ware and
Frank Cattle, pillars of our living–in community.

There was a healthy rivalry between Heany and Thornhill, which
if anything, intensified after Thornhill's switch in role. This rivalry
normally found an outlet on the sports field and running track, and
one of my first away games with the station rugby team was against
Thornhill. What a drubbing we took that day, thanks mainly to Mel
Channer, their gifted stand-off three-quarter.

What a pleasant change it was, when I had found my way into
the RATG side, to see the inspired use Mel made of the ball we
forwards had won. Channer's natural athleticism almost made the
rest of us despair, considering the scant regard he had for accepted
keep-fit discipline. Most of us found it necessary to give up smoking
completely during the rugby season, just in order to last through
the eighty minutes of play. Whereas, immediately before we ran on
to the field in the RATG games, Mel would be dragging heavily

on a cigarette as he briefed us on tactics. Seconds later, he was zig zagging his way through the opposition leaving them flat footed.

Once in a while, a Heany cadet came close to matching Channer's standard and, like Mel, gave us forwards real inspiration to win the ball. One such man was 'Straw' Hall. But such finds were with us only briefly. Then we were back to winning the ball for a stand-off who was a mere mortal like ourselves.

Occasionally, one of the units would upstage the other in the air. During Andrew Geddes's tenure, we achieved a memorable piece of one-upmanship by visiting Thornhill with a twenty-seven aircraft formation: Andrew personally leading this 'Balbo', and emphasising his Celtic roots with the R/T callsign 'Scotchclad Leader'.

Just to rub in our professionalism that day, we landed at Thornhill in a single formation. It must have looked extremely impressive from the control tower, where I imagine Bob Graham, the station commander, was standing. But in the aircraft on the extreme left-hand edge of the formation, I was in two minds just before touch-down as to whether or not to go round again, as I calculated what lateral clearance I had to my left. Fortunately for my career I did not have to overshoot: something which would have shattered the spectacle.

Between student courses, both Thornhill and Heany flexed their muscles with formation flights to Durban in Natal. We enjoyed good relationships with the South African Air Force (SAAF), and these exercises combined a pleasant recreational break with the discipline required to operate near to the limit of the Harvard's range.

En route, we stayed overnight at Zwartkop – the SAAF airfield at Pretoria – where we cleared customs, refuelled and also made a meticulous check on the weather. The closing stages of our route to Durban skirted the spectacular Maloti mountains of Basutoland: some of their snow-clad peaks rising above eleven thousand feet. And since the Harvard cruised well below that height, it was important on that leg to stay reasonably clear of cloud in order to keep track of our position.

At Durban, we stayed in the Garrison Officers' Mess: a well run, friendly establishment, close to the shore; where most days the menu boasted freshly landed fish. The other local treat was the opportunity for some sea bathing: a treat indeed after months in land-locked Rhodesia. And how exotic it sounded in our letters home to say we'd bathed within the safety of a shark net.

The return journey – with no customs formalities to delay us at Zwartkop – was done in the course of a day. So that by the time we landed back at base, we were beginning to tire: all the more reason, of course, not to put one's feet up too soon. A point well illustrated by one of my colleagues as we entered our dispersal, after an otherwise model visit to Durban. Within yards of our flight offices, this unfortunate man put a wheel of his undercarriage into a storm ditch, and there he finished his sortie, complete with bent propeller and crumpled wing tip. Jack Holt, our formation leader, who by this time was watching from his office window, turned a very nasty colour.

Some time later, I'm afraid Thornhill managed to make that incident look almost like a non-event. Not being entirely au fait with the circumstances, I must not pretend to give a balanced picture of their misfortune. But after their formation broke up in bad weather on the leg between Zwartkop and Durban, at least one Harvard landed wheels up on the sea shore, as it ran out of fuel. No doubt, unforeseen circumstances had a hand in this sorry débâcle, but of one thing I am certain, there were some pretty red faces at the subsequent enquiry.

To revert to the RATG rugby team, many of us at my level saw it as the one real unifying influence between Heany and Thornhill: when you've rucked together for eighty minutes, the man alongside you from Thornhill ceases to be 'one of them'. And if the interest shown by AVM Norman Alison, our AOC, was anything to go by, the team gave focus to headquarters pride too as we took on some of the better Rhodesian sides. But the one person off the field who really breathed fire and enthusiasm into us, was Group Captain Gus Walker, the RATG Senior Air Staff Officer.

Gus, although quite a short, lightly built man, radiated personality. And he had that rare knack of instantly recalling one's name, perhaps from a first meeting weeks before. This, and his ability to switch into completely unaffected conversation at any level, had us eating out of his hand in no time. With Gus on the touchline, we frequently played well above ourselves. The fact that he had played rugby for England of course added to his credibility in his role as unofficial coach.

Gus was also a darling of the crowd at the Bulawayo rugby ground, where he often refereed. During the war he lost an arm while attempting to rescue a crew from a burning bomber, and the extrovert sports-loving Rhodesians found the sight of this trim, one-armed referee (wearing his England jersey) very much in their own image.

Big Game

In our day-to-day flying we saw little of the wild animal life below. The exception being the occasional sighting of ostrich – invariably in pairs – when we were practising low flying. If we flew towards them, they would turn tail and bound away with their curious bobbing gait. But just before we overtook them, the male bird would turn about to confront the aircraft; its neck thrust forward and its wings beating furiously in a great display of anger as it sought to protect its mate.

From the ostrich's viewpoint, I can only imagine that the sight of a huge yellow bird suddenly swooping on it – a loud buzzing noise coming from its beak and with the wind whistling through its outstretched wings – must have presented a threatening spectacle. Whoever spread the story that the ostrich buries its head in the sand at the first hint of trouble, was short on fact. In spite of its faintly ridiculous appearance, with its long bare legs and scraggy neck, it seemed to me to show extraordinary courage.

Fortunately, to build on these chance sightings, there was a game reserve within easy motoring distance of Heany, where we could observe wild animal life on a more ordered basis.

My first visit to the Wankie game reserve was over a long weekend in 1950. It was an all-bachelor affair. David Ware and John Smith-Carrington supplied the transport and did the driving, whilst Jack Holt, Tony Young, Frank Cattle and I spread ourselves between their two cars as passengers, cooks and general hangers-on. Thanks to a helpful Mess steward, we set out self-sufficient in cooking utensils and rations. And a bulk purchase from the Mess bar insured against withdrawal symptoms.

In the course of that visit, with the exception of our attempts to photograph a rhinoceros, we heeded the notices warning visitors to remain in their cars. But on a visit a year later, I'm afraid the old adage of familiarity breeding contempt was well borne out. By then I had a car of my own, and felt perhaps that if I wished to take a chance, I could do so without embarrassing others. With Jack Holt as my passenger, I had just rounded a bend when we spotted the corpse of a baby elephant lying only twenty or thirty paces off the road.

A minute or two later, as we stood over the animal taking photographs, we realised that it was a fresh kill. More ominous, its stomach was a gory mess where something had only recently torn it away. Suddenly, I felt exposed and vulnerable, and those twenty or thirty paces to the car seemed a long way. From the look Jack gave me at that moment, he was having similar thoughts.

We cut short our photography and began to walk towards the car. As we did so a series of growls went up from behind the scrub nearby. My flesh tingled as I realised that we were virtually surrounded by a pride of lions. These animals were mere yards away, and had almost certainly been feeding on the carcass as my car approached. I think we were too stunned to break into a run, but once in the car we sat gasping with relief and muttering expletives non-stop for at least a minute. When we did resume a coherent conversation, neither of us was in any doubt – it had been a close call.

What makes that incident so foolhardy is the poignant reminder we had received, only the year before, of how dangerous the lion can be. During the return journey from my first visit to Wankie, we had called briefly on a cattle farmer who had been mauled by a lion. Tony Coombes was a tall, handsome, powerfully built man who looked the epitome of commonsense and self-reliance as he received us with a welcoming drink. But there was no escaping the fact that he had a false leg. For all his gumption, he had been no match for the lion's cunning.

As we sipped our drink and with a little persuasion, Tony had given us a matter-of-fact account of what had happened. He had set out with his African assistant to kill a lioness which was taking his stock. He was properly armed and was looking for the lioness when it pounced, having stalked *them*. Knocked down and separated from his rifle, as the lioness mauled him and he tried desperately to choke her by ramming his fist down her throat, he shouted instructions to his assistant, who grabbed the rifle and shot the animal. Tony could count himself lucky that he'd lost no more than a leg.

What Jack Holt and I had lost from that anecdote over the intervening twelve months, was not Tony's coolness or the courage and loyalty of his assistant, but the fundamental point that in the lion one is dealing with an animal whose basic instinct is to stalk and kill: something it has to succeed in week in and week out in order to survive.

Lunch at a Cattle Ranch

During the 1951 Christmas festivities, I received an invitation to lunch with Major George Errington and his wife who raised cattle on a ranch near Fig Tree, a few miles out of Bulawayo. I had been there some months before, and accepted without hesitation. The Erringtons were a cultivated, amusing couple who enjoyed good food and wine. Squadron Leader Donald Clause, our Mess president, was a fellow guest.

George, urbane to his fingertips, had served in tanks and had moved his family and possessions to Southern Rhodesia a year or so after being demobbed in 1945. He made no bones about how they had come to settle in Rhodesia. They had taken an atlas and shortlisted those places which offered a combination of – English as the standard language; a sunny climate; a plentiful supply of servants; and political stability. On that basis, Southern Rhodesia had selected itself.

And here they were, miles out in the bush, but eating regularly off the family dinner service, and surrounded by their choice pieces of silver, glass, furniture and pictures. They also dressed for dinner. And they managed all this without appearing in the least bit stuffy. George seemed equally at ease, whether he was pottering about the ranch in his Volkswagen Beetle in a faded shirt, corduroy trousers and suede 'brothel creepers', or impeccably dressed in a black tie.

Lunch that day was every bit as good as I had expected, and stylishly served too, with matching salad plates at each setting, a nicely chilled white wine with the fish course, and so on.

After coffee – well into the afternoon – George invited us to witness a little domestic ritual in which he was to present each of his farmhands with a token Christmas gift. With his staff assembled in front of the house, his head man called forward each hand in turn to receive his present from George in person. The gifts were identical – a boy scout's belt and an all-purpose clasp knife to hang from one of the belt's side rings. It was a simple but charming ceremony. Each recipient accepting his gift, making a slight bow and – in most cases grinning sheepishly – rejoining his smiling family and friends.

Towards the end of this event there was a brief exchange between the head man and the boy he had called forward, and George was informed that twice that day the boy had seen a leopard on a *kopje* (small hill) near to where the cattle were grazing. It was agreed that it would have to be dealt with.

George's armoury was limited to two shot guns – a twelve-bore and a four-ten. And in spite of Donald Clause's enthusiasm to tackle the leopard with the twelve-bore, George decreed that the situation demanded superior firepower and phoned the local police for assistance.

Twenty minutes later, George, Donald and I linked up with the police contingent – a European sergeant and an African constable – at the foot of the *kopje*: a granite outcrop dotted with vegetation and rising some two hundred feet above the surrounding land. Somewhere in the nooks and crannies of that boulder-strewn *kopje*, was the leopard; assuming it had not moved on.

George and the police sergeant discussed tactics. The latter and his assistant would climb the *kopje* armed with a .303 rifle. And

Donald, who with George's acquiescence had claimed the twelve-bore, would skirt one of the flanks in case the animal broke cover in that direction. At this juncture, as I took up the four-ten, I was instantly ordered by George to stay with him by the car. As he remarked, my weapon would be about as much use as a pea shooter against a leopard. I can't say I was really disappointed at being left out of the action: not as my mind went back to Tony Coombes and his encounter with the lioness.

About five minutes after the sergeant and the constable had disappeared among the boulders and vegetation, and begun their cautious ascent, we heard the crack of the .303. A single shot, followed by silence. No whoops of delight signifying success. No cries from a wounded animal. Nothing. We could only strain our ears and wait.

A few minutes later, Donald appeared. He had seen nothing and had returned merely to get an update on tactics. He seemed a bit put out when George said he should stay with us, pending the return of either the sergeant or his constable. But George quietly insisted, and Donald remained.

It was almost ten minutes after we had heard the shot, that the policemen returned, grinning broadly and obviously very pleased with themselves. The leopard was dead. They had flushed it from its hiding place, and the sergeant had shot it from the top of the *kopje* as it had broken cover at ground level. However, measuring something in excess of six feet – tip of nose to tip of tail – it had been more than they could carry for any distance.

After our return to the house, and the animal had been recovered with George's Land-Rover, we gathered round to admire it. It was a leopardess in cub. With its proud face and distinctive markings, it was a truly handsome sight. And for one, I could not resist stroking its sleek coat. The African farmhands and their families, however, although curious to view it from a few yards, absolutely refused to touch it: their timidity seemingly based on some taboo. These were Matabele Africans, and as far as I am aware, not a warrior race in quite the same mould as the Zulus. Presumably the latter would not have shown such inhibitions, since their shields are traditionally adorned with the skins of predatory animals.

That evening, Donald and I returned to Heany well wined and dined, and armed with a small story to slip into the conversation given a suitable moment.

OC Troops

The fact that I was appointed OC Troops for the journey home had more to do with my being a bachelor – without a wife to look after – than it had to do with seniority in rank. But I did not protest too much, since, with only two or three dozen service personnel on board, I saw the post rather in the nature of a sinecure than one involving real work. How wrong I was.

Two days out of Cape Town, one of my airmen tried to hang himself and had to be placed in protective custody. As the journey progressed, his mental condition was to occupy an increasing proportion of my day. To see him gradually deteriorate, until he was finally taken off at Southampton in a straitjacket, was a most depressing business. Nothing the ship's doctor could prescribe or I could say in the way of counselling seemed to make the slightest impact on his determination to do away with himself. A move to give him exercise on deck, suitably escorted, had simply resulted in him making a dash for the ship's rail. After that he was kept in confinement.

According to him, his Cape-coloured wife had married him merely to get a passport out of South Africa. When he discovered this, on the second day at sea, his reaction was to beat her up and to throw her passport through the porthole. Then, in an apparent mixture of humiliation and remorse, he had locked himself in their cabin and strung himself up to the overhead ventilation system. On the face of it, he had been well and truly taken in, and the sudden realisation of it had been more than he could take.

During the voyage, I had no option but to leave the physical handling of this harrowing affair to the ship's authorities: in matters of discipline, to the Master at Arms; and in the medical sphere to the ship's doctor. For their part they kept me abreast of developments; consulted me where it was proper to do so; and gave me regular access to the airman. For my part, I did my best to raise his spirits and opened a diary on the subject.

A few days out of Southampton, I saw it as my duty to alert the RAF authorities on shore to the problem coming their way. However, the Master at Arms informed me that the Captain saw it differently, and could not have me sending signals on my own account. I asked to see the captain, and at the end of a five-minute interview, he agreed to send my message virtually as it stood, provided it went out over his name. The job was done, with honour satisfied.

Within minutes of docking, an RAF officer from the shore establishment was in my cabin and I was handing over a copy of my diary. The airman was taken ashore and that is the last I have seen or heard of him. But I have often wondered if he managed to rebuild his shattered life. To complicate matters, his wife was some months pregnant at the time.

Having shed this depressing responsibility, it was almost a relief to do battle with the authorities over importing my car and obtaining a UK driving licence. Inside a couple of hours I was heading north west on the A36, eager to be reunited with my family and to show off my first motor car.

At home, where I was given an enthusiastic welcome, I was concerned to see that my father had begun to show his years: being not only much greyer, but noticeably less robust. Having said her welcome and approved of her present, my sister Sheila seemed more preoccupied with how I would manage without a garage: a fair point perhaps, but touchingly serious, coming from a thirteen-year-old.

Central Flying School

It was mid-October when I disembarked from the RMMV *Cape Town Castle*, and with something like six weeks leave in hand, winter had arrived before I set out for the Central Flying School (CFS) at RAF Little Rissington.

In the last few miles of my journey, as I climbed higher into the Cotswold Hills, a subtle change was taking place in the road surface, and one which I was slow to recognise. Suddenly, as I entered a sharp left-hander, the car acquired a mind of its own and carried straight on out of control. Fortunately I had the road to myself and went harmlessly through an open gateway into the field opposite. Underneath the loose powdery snow, the road surface was like glass, and as I resumed my journey, I was reminded that winter in a sheltered vale in Somerset is one thing, and on the higher reaches of the Cotswolds quite another.

Perhaps my posting to the CFS staff was an inevitable result of my upgrading to an A1 instructor's category. But had I been given the choice – assuming I was to remain an instructor – I would have opted to serve at one of the Advanced Flying Schools, where I would have flown the Meteor. As it was, I was stuck for the time being with my old friend the Harvard.

A nice feature of the Services is that one is frequently bumping into old friends. And so it was on my arrival at CFS. Two old friends from Heany were already well established on the staff. One with the Examining Wing, the other – perhaps less happily – in the post of Adjutant to the Station Commander. Shortly after his arrival, the latter had had the misfortune to stall a Vampire as he approached to land, which left the fuselage in need of a major rebuild. And since the Adjutant's chair became vacant at about the same time, you could say that in that moment of inattention he had unwittingly selected himself for the job. But my friend is not easily put down, and outwardly his self-assurance seemed unimpaired.

Although I was less than overjoyed with remaining on the Harvard, there were compensations in being posted to RAF Little Rissington. Principal among these was the presence of Group Captain Bill Coles, the Chief Instructor; a big, bluff, ex-policeman with a distinguished war record, who led from the front and whose stock-in-trade was a refreshing blend of pragmatism and fair play. Then there was Squadron Leader Leonard Trent VC, my squadron commander; a New Zealander – whose decoration speaks for itself – and a thorough gentleman. With both of these I registered my wish to move on to jets, and having done so, got on with the job.

CFS comprised two parts – the Flying Wing and the Examining Wing. The former included two Harvard squadrons and a Meteor squadron, whose respective functions were to train flying instructors to staff the Flying Training Schools (FTSs) and the Advanced Flying Schools (AFSs). The Flying Wing also had a flight of DH Vampire aircraft which operated separately from the Meteor squadron; initially equipped with the single-seat FB5, which was joined later by the dual-seat T11. Other than the use of a communications aircraft to discharge its roving commission, the Examining Wing had no aircraft of its own. Its function within Flying Training Command was to monitor the standard of instruction at the FTSs and AFSs, to recategorise the instructors as they gained experience, and to appoint Instrument Rating Examiners. The Examining Wing also had responsibilities to the RAF operational commands and to Royal Navy and Army flying units; responsibilities which I touch on later.

A unit with an examining role can hardly expect universal popularity: the best it can hope for is a reputation for objectivity and professionalism. And whatever it does, it is going to be the target for some sharp digs from time to time. Hence the claim of one wit that the pelican on CFS's armorial bearings was appropriate because it has a big mouth and doesn't fly very well. Not intended as an objective summing up I'm sure, but deliciously acidic.

The ratio of students to instructors in the CFS Harvard squadrons was noticeably less than it had been at Heany. Even so, with the vastly inferior weather factor at Little Rissington, there were periods when we had to put in much longer hours to get the courses out on time. This was particularly so in winter, when we sometimes worked right through the weekend. No one minds an occasional weekend's work, but when you get two or three in succession, even the crew-room comedian grows quiet.

This winter grind meant that the arrival of spring was greeted with especial enthusiasm at CFS. How good it was to return our snow shovels to stores and to put the (runway) snow-clearance plan back in its file; and also to see the back of those novel attempts to get rid of the more stubborn icy patches with the jet efflux of our Vampire, as it was taxied into position like some enormous blowlamp. How good it was too, to see patches of blue sky and sunshine without having to climb through two or three thousand feet of solid cloud in order to do so. After a winter on the hill (750 feet above sea level) – in fog which sometimes persisted for days – the Scandinavians' habit of pouring south in the spring no longer seemed in the least over-the-top.

A few weeks after joining CFS, I was offered a sortie in a Vampire – the FB5 single-seat version. An aircraft kept primarily to give our Harvard students a smattering of jet experience before they joined their FTSs. With no dual seat version available at that time, it was a matter of studying the Pilots' Notes and being briefed by someone who had flown it. Flight Lieutenant Peter Middleton, who despatched me on that initial flight, kindly warned me of the stony silence which would follow as I pressurised the cabin after take-off; thereby saving me some unnecessary palpitations as the engine appeared to cut out. It was a thoroughly enjoyable flight. The Vampire hadn't the phenomenal rate of climb of the Meteor, it's true. But compared with the Harvard, it had plenty of fizz, and – having a pressurised cabin – it was even quieter than the Meteor.

That Vampire sortie merely sharpened my longing for a permanent switch to jets. And my lobbying to do so at last yielded a jet-familiarisation course. From the beginning of April (1953) I spent a fortnight at 205 AFS Middleton St. George, flying Meteor T7s and the single-seat Meteor F4.

Waiting to greet me on my arrival at the flight line was one of my FTS students, Sergeant Waring, who informed me with a cocky smile that he had specially asked for me as one of his students. Well, if his intention was to get some of his own back, he at least did it without malice. Yes, he worked me hard. Much as I had him, I'm sure. And he could not resist remarking that *results count, not excuses*, as I sought to analyse my shortcomings in one of our post-flight discussions: a phrase I had been fond of at Heany apparently.

But for all Waring's obvious satisfaction in our switch of roles, it was a tonic to fly with him; his technique had such elan. And it was impossible not to feel some small degree of personal satisfaction with the standard he had achieved. Certainly, in flying the Meteor, he set an enviable standard.

But how slowly one accumulated flying time on those short-range

ABOVE. RAF Heaney 1st fifteen, 1950. LEFT. Ken and Moyca are married *(Greytown, Natal, 1950)*.

ABOVE. CFS Examining wing keeping its hand in, 1954. *left to right:* Sabre F4, Venom FB1, Hunter F1, Canberra B2, Meteor T7, Meteor NF14, Vampire T11. RIGHT. CFS Aerobatic team, 1954 led by Flt Lt Bob Price *(Photo's courtesy of Avro International – Ian Lowe).*

jets. With sorties averaging only forty minutes or so, the build-up was even slower than with the Harvard. The seventeen sorties of my familiarisation course amounted to fewer than thirteen hours: less than *one* of my longer Liberator sorties. As for sheer stimulation however (wartime operational factors apart), there was no competition. Flying the short-range jet won hands down. I'm afraid that returning to Rissington and the Harvard was quite an anti-climax.

Summer 1953, with the Queen's Coronation, brought a welcome touch of colour to our lives, as well as keeping us more on our toes. In mid-June, the Mess held a Coronation Ball, a glittering affair, with lots of senior RAF guests, as well as important people from the local community. By this time I was well installed as the messing member, and under the leadership of our president, Wing Commander Bill Renwick, I shared the onus of the Mess Committee generally of seeing that the ball arrangements went smoothly and, more importantly, that they were seen to do so.

Thanks mainly to our Mess staff, headed by stalwarts like Mr Grey, the Mess steward, and Jack Stevens, our unflappable Mess secretary, our guests voted it a great success. As for the Committee: apart from conveying these sentiments to the staff, our chief reaction was to heave a sigh of relief. In my less responsible moments, I've sometimes wondered what noticeable difference it would have made to the outcome of such functions had the Mess Committee not existed. But then, if things *had* gone wrong in those circumstances, who would there be to walk the plank?

The following month, the Queen reviewed the RAF at Odiham: the principal event being a flypast. Given the need to involve the aircraft of each of the commands – with their wide diversity of bases and aircraft speeds – there was an obvious need for thorough rehearsal. The last thing Her Majesty would want to see would be the jets overtaking the propeller-driven aircraft in an unseemly gaggle, so rehearsals began weeks before the event.

The other part of the review was a large static aircraft display: the machines lined up, wing tip to wing tip, row upon row, with their air and ground crews in attendance. It was in this lesser role that I took part. But again, nothing was left to chance. My colleagues and I had the CFS contribution (comprising three Harvards) in position three whole weeks before the event. And being obliged to remain at Odiham from that point onwards, by the time Review Day arrived, we had seen the flypast progress from the occasional shambles to synchronised perfection. Come the day, we had grown just as proud of the RAF as the monarch herself.

Since no campaign medal was struck for those of us obliged to live under canvas during those three weeks of waiting, we did the next best thing; commissioning a special tie from Gieves. This was liberally embellished with earwigs; symbolic of the real thing which had invaded our tents in their scores, and had nipped us in our bed rolls night after night, regardless of how thoroughly we cleared them out before turning in. What luxury it was to return to Rissington and to wake up without squashed earwigs adorning our torsos.

Life on 4 (Harvard) Squadron was not all diligent application. It had its lighter, some might think irresponsible, moments. Shortly before converting to jet instruction, and during a break between courses, I flew to Exeter; a mere thirty miles beyond Wellington, my home town. The idea being to keep the polish on my navigation and at the same time to give one of our ground crew a day out. But in choosing our destination, and my passenger come to that, I must admit to having something additional in mind. Which was why I particularly asked for a passenger with a strong stomach.

It was a sunny day with well scattered cloud. And with no retired Air Marshals living within twenty miles of Wellington as far as I knew, I felt it was high time the citizens of that fair town shared – however modestly – in the delights of an aerobatic display. No beat-ups at tree top height, but something around two or three thousand feet and safely displaced from the town centre over open countryside. By then I had accumulated close to a thousand hours in the Harvard, and flying it had become almost second nature. As I saw it, the circumstances might never be more opportune for showing off a bit to one's friends.

After a light lunch and refuelling at Exeter airport, I made for Wellington. Conditions could not have been better. The timing too was auspicious, with people returning to their offices and factories after lunch. I put the Harvard through its paces. Five minutes or so of continuous manoeuvre, finishing up by flying inverted until the engine coughed and spluttered in its usual dramatic fashion. It all went very smoothly. And my passenger at least, as well as hanging on to his lunch, seemed quite impressed. But how had it looked, I wondered, from the ground? Well, if it had seemed tame or miles away to the Wellingtonians, I need not claim responsibility.

Later that day, when I motored home for the weekend, I got their reaction. Generally, they seemed most impressed, and I received a number of complimentary remarks from friends and acquaintances in the Sanford Arms that night: not least from Harold Hine, the landlord, who was especially taken with the inverted bit at the end – the least difficult part, as it happened.

My father's reaction, however, was rather different. He told me he had watched the whole thing, and that on being asked by several people if that was his son putting on the display, he had replied that his son would not be such a bloody fool as to misbehave in that manner, especially over his home town: I don't recollect him passing an opinion on the quality of the display one way or the other. Retired Air Marshals or not, I'd received my ticking off.

The opportunity to switch to jet instruction did not come until almost five months after my jet familiarisation course at Middleton St. George, by which time I was more than ready for it. In August 1953, I reverted to the role of trainee instructor and began a CFS jet course on the Meteor T7.

One of my instructors was Bob Bragg, a twenty-one-year-old ex-Cranwell cadet. Bob, who had had no front-line squadron experience, was among a handful of ex-Cranwell cadets to come directly to CFS at the end of their AFS training, and who joined the CFS staff directly they had completed their instructor's course. This method of staff recruitment was in the nature of an experiment and was less than ideal perhaps, since those concerned had to handle students of much greater maturity than themselves. Nevertheless they coped remarkably well, and there was no questioning their above-average ability as pilots. And what they lacked in operational background, they largely offset with their zestful approach and a readiness to challenge accepted doctrine. A readiness which occasionally irritated some of their older colleagues, but which ensured that none of us fell victim to complacency.

The Meteor instructor's course dealt extensively with action in the event of engine failure: much of the practical application being devoted to overshooting and landing on one engine. To begin with, we merely throttled back one of the engines; which meant that the 'dead' engine could always be reintroduced if one had seriously misjudged the approach. But as the course proceeded, we increasingly practised these exercises with one engine completely shut down; from which there was no retreat. Once below six hundred feet in that situation, we were committed to land. And since at the point of decision one was still quite some distance from the runway threshold, the exercise never entirely lost its ginger.

Given the general level of experience among the CFS staff and students, this policy of single-engined training gave rise to no more than one fatal accident at Little Rissington that I can recall. But at the AFSs it produced a rash of such accidents which, in the mid-fifties, led to a general edict from the Air Ministry forbidding the shutting down of engines on the Meteor below five thousand feet, except in an actual emergency.

This decision must have been a difficult one, since in a way it could be seen as running away from the problem, and an abandonment of the CFS doctrine of flying the aircraft to its limits. Indeed, it was said that the Chief of Air Staff himself (Sir Dermot Boyle) had had the final word in the affair. As a former A1 CFS instructor, Sir Dermot probably agonised over the decision. But his staff's reasoning was one of hard-headed reality: namely, that actual engine failures in the Meteor were so rare, they were accounting for only a fraction of those killed flying the aircraft on one engine. So that, in effect, the cure was proving infinitely more costly than the problem. Subsequently, all single-engine practise below five thousand feet was done with the 'dead' engine merely throttled back to idling power.

A few months after I had transferred to the CFS Meteor squadron under the command of Squadron Leader Alex Harkness, Alex himself was faced with the real thing, and at the worst possible moment:

the sudden and complete failure of one engine just after lift-off from a rolling touchdown at night.

A big man with a powerful physique, Alex needed all his strength and that of his student too to maintain directional control, as they struggled to gain height on their remaining engine. But for the fact that Rissington is on a hill, they would surely have been obliged to crash land straight ahead, with death or serious injury very much on the cards. However, thanks to the valley at the end of the runway, they were able to build up vital speed before attempting to regain circuit height. Even so, it required a superhuman effort on the rudder bar, and a great deal of skill, for Alex and Peter Stonham, his student, to complete a circuit and to land safely on the runway. But they did just that. An experience which cost Alex a few grey hairs I dare say, but one which gained him an immediate bar to his Air Force Cross.

The Meteor T7, with its unpressurised cabin, but able to fly to 40,000 feet and beyond, and with its exceptional rate of climb and descent – imposed on us altogether new physiological stresses. And whilst medium-level sorties could be taken more or less in one's stride, three or four high-level sorties in a day had us flopping on our beds after tea and immediately falling into a deep sleep. Without being particularly conscious of it at the time, we were part of the learning process in determining the safe limits of operation for those early, unpressurised jets. A few years later, a limit in the region of 30,000 feet was laid down for the T7. Meantime, we were more or less free to explore the limits for ourselves.

One such exploration took place on a staff-training visit to Malta, during which our chief flying instructor, Wing Commander 'Fearless' Frank Dodd, led four of us on two high-level mock attacks on Valetta Grand Harbour. The first of these was made at 35,000 feet. But, by way of putting one over on the defending fighters (Vampires), on our second sortie Frank brought us in at 40,000 feet, and we had completed our attack several minutes before the Vampires made their interception. (Maintaining close formation at that altitude at a speed of .72 Mach was no easy matter, with the throttles already fairly wide open, and I eventually settled for leaving the throttles set and using the airbrakes as a speed control. It might have contravened orthodox engine handling, but it worked beautifully for the short period involved.)

After some minutes at 40,000 feet on that second attack, I was aware of a strange tingling sensation in my wrists. I was beginning to suffer the effects of decompression. And more effects were to come. Following our rapid descent to Luqa airfield, my passenger (a pilot from the locally based Shackleton squadron), rushed to the grass verging our dispersal and heaved up his breakfast. I was suitably apologetic for ruining his first jet flight in this manner, but what I did not tell him was that I was not feeling too well myself.

Changing to a more sombre note: as the vast majority of fatal flying accidents occur miles from the aircraft's base – usually in open country – most occupants of RAF stations are shielded from the grim spectacle of the crash itself. So that when a fatal crash takes place on the station the effect can be equally as shocking as it would be to ordinary members of the public. Such was the case when a Meteor crashed within yards of our officers' married quarters, killing the pilot and hurling debris in all directions.

When the crash occurred, I was closeted in a nearby sports changing room waiting for the arrival of a visiting rugby team and, although the remains were scattered within a hundred yards of us, it was a minute or so before we realised what had happened. As we emerged, the scene was a mixture of hectic activity, twisted wreckage and small groups of stunned onlookers. Our medical staff were working like beavers to deal with those injured by the flying debris and to collect the remains of the pilot, whilst the onlookers slowly dispersed as they realised they had nothing to contribute. Needless to say, we played no rugby that afternoon.

The pilot had been engaged in an end-of-course aerobatic competition when, for some reason, he had lost control. It would be wrong to think that such matters are soon forgotten – indeed that event is as horrendous and sad to me some forty years later as it was that very afternoon – but flying is the business of the RAF, and after burying our comrades with honour and respect, we have to put these sad occasions to one side and get on with the job.

To return to the job of the Examining Wing: whilst its prime function was to examine flying instructors and students, the Wing also had responsibilities for pure flying standards within the RAF operational commands – responsibilities which also applied to the shore-based flying units of the Royal Navy. It was this link with the operational commands which gave particular interest to the job, since it kept members of the Wing in regular flying practice on front-line aircraft.

This link entailed much more than the monitoring of general flying standards. For instance, during the period involved, the Wing was asked to give practical advice on how to identify and recover from an inadvertent spin in high-performance aircraft: a predicament which the fighter pilot occasionally encountered as he flew his aircraft to the limits of manoeuvre during practice combat. With that sort of responsibility, members of the Examining Wing – working in conjunction with the Handling Squadron at Boscombe Down – occasionally found themselves pushing at the limits of pilot handling knowledge, as they put their theories to the practical test.

Whether such testing contributed to the accident when two members of the Wing decided to abandon a Sea Fury I can't say, but my understanding is that they were practising spin recoveries at the time. One of them parachuted to safety, but the other crashed with the aircraft and was killed. As the officer despatched to mount guard over the crash, the only consolation I could draw from the scene was that death for the one trapped in the aircraft must have been mercifully swift. As with the young man who died in the aerobatic competition, it was a tragic business, but flying at that demanding level inevitably carries additional risk.

A series of spinning incidents, involving both the training and operational commands, made spinning the subject of avid discussion in the Flying Wing crewrooms at that time. And although our Meteor syllabus included practice spin recoveries as a routine exercise, those dealt exclusively with *erect* spinning – with the aircraft the right way up – whereas at least some of the incidents appeared to concern spins with the aircraft inverted. It was this latter variety which intrigued us, because, in that situation, the visual and instrument indications of the direction of spin were said to contradict one another: a fundamental difference from the erect spin, where they are in harmony. However, since inverted spinning was banned as an intentional manoeuvre, we could only speculate on the problem.

At least that was the position until our Examining Wing brethren passed on to us the outcome of some Meteor inverted spinning trials, which included a proven method of recovery. With that, I'm afraid the temptation to see for myself proved irresistible.

Joined by Flying Officer Charlie Slade, who got wind of what I was up to, I climbed to what I felt was a safe starting height and began the usual pre-spinning checks; only to have Charlie ask if I would mind starting from a bit higher. This smacked a bit of procrastination to me, and anyway I fancied our recovery action would take effect that much quicker in the thicker air lower down, but I agreed, and up we went another five thousand feet or so.

This time I knew that if I did not get on with it, I might myself begin to have reservations, so in we went. With the aircraft inverted and approaching the stall, I eased the stick progressively forward and at the first hint of the stall, applied a bootful of rudder. A moment's hesitation and the aircraft slowly entered an inverted spin. So far so good. Charlie and I read the instrument indications aloud and agreed that whereas the turn indicator showed us going one way, from visual indications we appeared to be rolling in the opposite direction: the indications we had been told to expect. By now we had lost about four thousand feet and it was time to take recovery action.

With Charlie following through on the controls, I applied full rudder against the turn – as shown on the flight instrument panel – and smoothly eased back on the stick. As we neared the vertical, and the axis of the spin steepened, so our speed of rotation smartly increased. Just then the rudder began to bite, and the next moment we had stopped the rotation, centralised the rudder and were easing out of a steep dive. All this, I might say, to considerable relief and satisfaction from both cockpits.

Whether we did a second inverted spin, in the opposite direction, with Charlie operating the controls, I can't remember: we probably did. But what I do recall was the disconcerted look on Alex Harkness's face when we let slip what we'd done. He was quick to tell us that a decision had been taken not to practise inverted spin recovery within the command. Well, it was news to Charlie and myself that a positive edict had been issued, but we made a suitable apology and undertook not to meddle with it further.

Once at home on the Meteor, I put together a solo aerobatic display. Something lasting about four to five minutes, and which was not too dependent on a high cloud base. Other than the fact that it opened with a high speed arrival, it was similar to my Harvard sequence, and embraced Cuban eights (two-thirds of a loop, followed by a half roll – repeated), Derry turns (rolling inwards during a steep turn, to turn steeply in the opposite direction), an assortment of rolls and some inverted flying. Things known in the trade as 'no sweat' manoeuvres, where the risk of losing height inadvertently is kept to a minimum.

Practising my sequence at Rissington, I received a polite but firm request from Air Commodore Christopher Paul, the commandant, to pay closer attention to the rules governing minimum height, so I consciously tamed it before putting it on at the 1954 Baginton air display. In the event, at least one of my colleagues thought I had tamed it rather too much; but to judge from the next edition of the *Flight* magazine, it at least passed muster with the aviation Press.

My appearance at Baginton was not without incident, I regret to say. I was scheduled to give two displays; one on each of two successive days. On the first occasion, I lowered the undercarriage somewhat too soon in an effort to reduce speed more rapidly, with the result that one leg became jammed in the partially extended position; forcing me to abandon the display before I was properly into it. I diverted to nearby Bitterswell, landing on the grass crash strip alongside the runway, making the best use I could of one main wheel and the nosewheel. With the extra drag from my partially extended main wheel, and given my specially reduced fuel load, I was not prepared to gamble on reaching Rissington.

The landing could hardly have gone better. After touchdown, I held the wings laterally level until I ran out of aileron control, and as the wing on the affected side finally scraped the ground, the aircraft slewed round onto the edge of the runway in a graceful arc, obligingly, remaining the right way up. Seconds later, as I stepped from the aircraft, without as much as a bruise or a scratch, I was bundled willy nilly into an ambulance and whisked off to the control tower to be checked over and debriefed.

Twenty-four hours later, I was back over Baginton in a replacement aircraft. On this occasion, I observed the undercarriage lowering speed with meticulous care, and my display went off as planned. The Court of Inquiry, set up at Rissington to investigate the incident of the previous day, attached no blame to my piloting, given the special latitude allowed display pilots in their handling of their aircraft. But I counted myself fortunate in being handled by a sympathetic Board. Maybe the fact that the aircraft – a distorted undercarriage jack apart – had suffered so little damage, did something to swing the outcome in my favour.

Although others emerged from time to time to give solo aerobatic displays, the undisputed king in that field was Flight Lieutenant 'Flush' Kendall. His display in the Vampire T11 was a beautifully precise and compact affair; full of variety and quite stunning to watch. The way Flush coolly half-bunted the T11 only a few hundred feet off the ground left everyone at all knowledgeable of what was involved open mouthed on seeing it for the first time. On the ground, the only evidence of this amiable man's aerobatic prowess was the bloodshot nature of his eyes: the result of all that negative 'g'. Unfortunately, 'Flush' was to die only a year or so later during air-to-ground firing practice whilst serving on a Sabre squadron in Germany. A cruel end, given his particular talent.

During 1954, Flush shared the top-level CFS aerobatic scene with a team of Meteor T7s led by Flight Lieutenant Bob Price: a man whose natural touch equipped him equally to shine at golf and cricket, and – on occasions – even with the billiard cue. Once or twice a week during the display season, Bob's formation of four opened the flying day by practising their sequence over the airfield: a necessary part of retaining their polish, but also useful in inspiring the rest of us, and showing our students that CFS staff could *do* as well as teach.

At the mention of formation aerobatics, the Red Arrows spring to most people's minds – and of course, their standard is a thing apart. But part-time teams, such as the one led by Bob Price, called for almost greater commitment in some respects, since most of their practice flights, as well as the displays themselves, took place outside their normal working day. Even so, there was nothing part-time about the quality of our Meteor team's display, and it was much in demand on the display circuit; as well as delighting the ageing members of the CFS Association as they stood on the Mess lawn before their annual dinner.

1954 an emotionally stressful year in my private life. My father was ailing from Christmas 1953 onwards with terminal cancer, and in spite of rallying after his first bout of surgery, he died in Frenchay hospital in late November. Fortunately, I was able to visit him regularly during his first stay in hospital. But during what proved to be his terminal phase in Frenchay, I was suddenly posted from CFS, and he died a week or so after I had reported to my new unit. Although I had known for some time what to expect, and rationally I had accepted the inevitable outcome, my bonding with my father was such that his death affected me for some time afterwards.

To return to my rather abrupt departure from CFS: it derived entirely from a chance meeting, some six months earlier, with an ex-25 Squadron pilot serving in the Air Secretary's Branch. He was attending a guest night at Rissington, and begged a Meteor flight to keep his hand in. And since he was out of practice, he had to be accompanied: by me it so happened.

Although basically a very capable pilot, Flight Lieutenant Roy Bowie recognised that his desk duties had left him a little rusty, and he pressed me to spare him nothing in re-acquainting him with single-engine flying. Here was a man with a professional approach, and I took him at this word. The harder I worked him the better he liked it. We began by re-establishing his critical speed (the lowest speed at which he could still maintain directional control under full asymmetric power) and then concentrated on single-engine over-shoots and landings. At the end of our forty-minute sortie, we were both bubbling with satisfaction: Bowie, because his underlying skill had quickly shone through, and me because it's always rewarding to meet such motivation and to have a hand in restoring talent.

Six months later, when the Commander of 25 All-Weather-Fighter (AWF) Squadron asked Bowie to find him a Flight Commander, he thought of me.

That was the background to a phone call from Bowie in late October 1954, asking me if I would accept the posting of flight commander designate 25 Squadron. The offer was conditional on my taking it up without delay and going direct to the squadron. Such offers do not grow on trees, and in spite of it distancing me from my father's sick bed, it was one I could hardly refuse.

Within days, I was settling in at RAF West Malling, having completely bypassed the normal routing through the AWF Operational Conversion Unit. When I queried the wisdom of taking this short cut, Bowie had said, 'look, 25 Squadron know your background and are happy to train you in all-weather tactics on the squadron; now do you feel up to taking the posting or not?' Put that way, I had no hesitation in saying I did.

10 *All-Weather Fighters*

When I arrived at West Malling, in November 1954, 25 Squadron shared the base with two other Meteor squadrons: 85, a sister AWF squadron commanded by Squadron Leader Basil Scandrett, and 500 County of Kent Royal Auxilliary Air Force Squadron equipped with single-seat day fighters and commanded by my squadron commander of 4 FTS days – Squadron Leader Donald Clause. Early the following year, our complement of AWF squadrons was brought up to three with the formation of 153 Squadron commanded initially by Squadron Leader Billy Veitch and also equipped with Meteors.

Squadron Leader Cameron Cox, the 25 Squadron commander, gave me a most cordial welcome. He had natural charm and an easy manner, but underneath one sensed a serious mind, to which standards mattered. We were quickly in discussion on how best to absorb me into my new role. It was almost ten years since I'd been on a front-line squadron, and it was my first time in fighter command; what's more, I'd bypassed the OCU. I had some catching up to do.

One thing heavily in my favour was that 25 Squadron was equipped with Meteors: Mk 12s and 14s it's true, but fundamentally Mk 7s with a radar tacked on the nose and two 20mm cannon installed in each wing. The net effect of this extra equipment was to rob the Night Fighter Meteors of much of the performance of the Mk 7, so the challenge facing me was not one of mastering the aircraft, but of learning to use it in the AWF Role. (The only way in which the AWF Meteors were superior to the Mk 7s as aircraft – apart from the Mk 14's superb all-perspex canopy – was that their cabins could be pressurised, thereby substantially increasing their safe operating ceiling.)

To succeed in adapting to my new role – my own efforts apart – I would have to make the most of other people's inputs. Not only that of my navigator/radar operator (Nav/Rad) but also of those controlling me from the ground: the fighter control network. Cameron immediately sent me off to make my number with the latter, and to get a feel for how *they* did things.

How coldly impersonal the blips of the target and the fighter aircraft looked on the radar screen of the Fighter Controller. And yet, here he was, not wasting words as though he was engaged in bar-room chat, no, but still managing to convey in his patter a sense of personal involvement; as though he understood the problems of manoeuvre at 35,000 feet and that to him it was as crucial that the interception should succeed as it was to the aircrew themselves. In time, I was to discover that some of the most effective controllers were women. Perhaps it had to do with that extra empathy in their voices.

During my return journey from the Ground Controlled Interception (GCI) station at RAF Bawdsey, I was informed of my father's death; and a few days later I attended his funeral. At least he was now out of pain. He had endured his illness with great courage, but what a grim business it had been to watch him suffer.

It was late November before I seriously tackled my new trade from the cockpit. After that the sorties came in plenty. Less than a fortnight later, during a night 'raid' by Canberras operating from bases in West Germany, my Nav/Rad and I had the good fortune to nail one of the attackers. It was a delicious moment to come up alongside our moonlit 'victim' at 35,000 feet and to announce a kill to our GCI controller. But, sadly, there was a phoney side to it. Had the Canberras not been throttled back, we wouldn't have stood a cat's chance against them: such was the obsolescent state of the Meteor against a modern bomber.

At that stage, the delta-wing Javelin – the Meteor's intended replacement – had still to enter service, but there were already rumours that its top speed performance, although better than the

Meteor, had not lived up to expectations, and that its introduction might not represent the great leap forward the RAF all-weather world was eagerly awaiting. One could hardly point the finger at anyone over the deficiencies of the Meteor. In one form and another, it had been in service for ten years and aircraft design had moved on. But if what was being said about the Javelin was true, the outlook for the AWF squadrons was hardly encouraging.

In fact, a series of accidents with the prototypes delayed the introduction of the Javelin until early 1956, when the Mk 1 first went into service with 46 Squadron at Odiham. 25 Squadron had to wait until the summer of 1960 before it received its Javelins – the Mk 9. By which time the fuel tankage had been extended and a number of engine and aerodynamic changes had been incorporated to improve performance and handling. Reflecting an objective as well an informed opinion on the Javelin is not for me the easiest thing, especially since my own experience of it is confined to a single familiarisation sortie. But, in the absence of a decision to develop a thinner wing version, I believe it did disappoint in terms of top speed. That is not to say that its speed advantage over the Meteor was not welcome, and it undoubtedly conveyed very distinct advantages over the Meteor in other areas, such as its service ceiling and its ability to achieve kills further out from the UK coastline.

The everyday pattern of our practice interceptions (PIs) was to take off in pairs; and once identified by our GCI station, to split up under the controller's direction, taking turns as fighter and target. It worked well enough: rather too well in fact, since it omitted the realism of how the fighter is despatched under operational conditions. During formal exercises this deficiency was removed by assembling our aircraft on the Operational Readiness Platform (ORP) – alongside the take-off point – from where we received the order to 'Scramble' (take-off) direct from the GCI sector controller. With an external battery cart already plugged in, engines were started in seconds, and within a minute or so we were in the air and turning on to our heading for the climb.

Crew rivalry to achieve the most rapid scramble added spice to the take-off phase, and both the ground and aircrews pulled out every stop in an attempt to beat the clock. On one occasion, with the adrenaline whipping him into action, one of our captains did not even wait for the ground crew to unplug the battery cart, before charging down the runway, dragging it with him. There's keenness for you.

Taking off on the first night flying detail, and climbing immediately to high altitude, the rays of the setting sun playing on decaying condensation trails, could sometimes produce weird effects. Whether or not this was in part responsible for what I am about to recall, I am still not certain. As part of the first detail, we were flying west one night at about 25,000 feet – the sun's rays still visible – when I saw what appeared to be a very fast-moving, brightly lit object, breaking up in the sky some miles ahead and between five thousand and ten thousand feet above our altitude.

Talking afterwards with the other crews on that first detail, they not only corroborated my sighting but added to its mysterious nature, so that I reported it up the line through the operations network. In the morning, a message was waiting for me from Group Captain Peter Hamley, the station commander – 'would I agree to appearing on TV that evening to answer questions on the sighting?' It seemed there was a mole in our operations net who had tipped off the media, and the BBC – regarding our sighting as a possible UFO (Unidentified Flying Object) – wanted to feature the story live on that evening's news spot.

Early that evening, having rehearsed my story with the interviewer (Mr X) at the Alexandra Palace TV studios, and whilst we waited for Dr Porter, an astronomer, to complete our discussion group, the producer looked in. At which point X suggested taking me out for a drink. 'Certainly,' said his producer, 'after the programme of course.' X's jaw dropped noticeably.

I'm in no sense doubting X's wish to be hospitable, but I think that between the two of us, it was probably he who was most in need of a pre-broadcast drink. To give him his due, he was very effective in winding us up for the interview and in handling the discussion itself. But to judge from the beads of sweat erupting on his face as we proceeded 'live', it was taking a lot out of him.

Afterwards, when we did get our drink, he explained that he was only on attachment to TV, and could hardly wait to get back to good old-fashioned radio.

As to the outcome of our discussion on TV, Dr Porter was unable to offer a convincing explanation of what we had seen in scientific terms. And since the national newspapers began a fairly lengthy strike the following day, neither did our sighting receive the benefit of an airing in the Press.

Enough of UFOs. While the perceived bomber threat was mainly one from high altitude, and high-level PIs were our bread and butter, once in a while we operated at the other end of the height band. On a dark night over the English Channel, with the target flying at six hundred feet and the fighter at four hundred feet, it called for a fair degree of concentration in the fighter role. And more than once, as we overflew the tall superstructure of the larger ships – their lights ablaze – I looked hard at my radio altimeter to double check that I really was flying at four hundred feet. Glancing below, those ships looked mighty close.

The striking thing about 25 Squadron 'B' Flight crew room was the youthfulness of its occupants. In my two years at CFS, mixing almost exclusively with sober-minded thirty-year-olds, I'd forgotten how exuberant and carefree men in their early twenties can be. How refreshing it was to recapture a little of their cheerful optimism and sense of fun. As their future flight commander, however, I could see that I might have to encourage a little more restraint in their attitude towards local council property: the crewroom and its approaches were littered with trophies from that source. The common denominator of this illicit collection being that each item could be loosely interpreted as referring to the work of 25 Squadron – the prize piece being a large red-on-white notice reading DANGER – MEN AT WORK OVERHEAD.

85 Squadron's living-in members chose to display their individualism more to the outside community. Their social transport, in which they attended local dances and the like was, unmistakably, a converted hearse. Their carefree attitude was summed up by the large notice hanging in the rear window which read DONT LAUGH AT THIS

VEHICLE MADAM – YOUR DAUGHTER IS PROBABLY INSIDE. The mothers might not have approved of the spirit underlying this gesture, but their daughters certainly did.

Intercepting a target is one thing; shooting it down, another. These were the days before fighters were equipped with heat-seeking missiles, which home on to the target automatically after release. We had to master the intricate business of the quarter attack. So, next to PIs, most of our time was devoted to air-to-air gunnery practice shooting at a strip of canvas towed by a tug aircraft. Opening fire at a wide angle, as one closed in on the target's quarter, demanded extreme accuracy of aim; narrowing the angle too much, and you were in danger of hitting the tug. Some, but very few, were natural shots; the rest of us worked like stink to get it right, without ever being able to guarantee a good result. In my flight, Flying Officer Fred Lundy was one of those naturals who could, it seemed, reel off a good score at will; how we envied him.

Real improvement in that old-fashioned variety of air-to-air firing was not achieved through infrequent dabbles off the nearest coastline: it required continuity. So it was no coincidence that each year every RAF fighter squadron put in a month at an Armament Practice Camp (APC). In the spring of 1955 we made our way to the APC at RAF Acklington on the Northumberland coast. It was a bracing experience: first, thanks to the fresh sea air, and next because of the know-how and the go-go attitude of the APC staff. As a squadron, we benefited considerably. As an individual, although my shooting had its moments – as when I scored a dozen hits on the flag, and the occasion when I chopped the wing off a 'glider' target and sent it spiralling into the sea – it showed that I still had something to prove.

Whilst we were at Acklington, Cameron went into hospital and I was left to run the show. Shortly after our return, the post of Squadron Commander was upgraded to the rank of Wing Commander, and Peter 'Jamie' Jamieson took over. Jamie was energetic, professional and rather intolerant of those who did not match up. My three months under his command were frequently stimulating, sometimes amusing, and never dull.

219 Squadron

In October 1955, on promotion to Squadron Leader, I took command of a flight on 219 AWF Squadron, a DH Venom NF2 Unit based at RAF Driffield in Yorkshire. At that stage 219 was still in the work-up phase, having re-formed only the month before, and our squadron commander, Wing Commander Ray Watts, had still to get the squadron on the Sector operations board. We shared the base with 33 Squadron, also equipped with Venoms, and commanded by Wing Commander Dick Patrick.

The Venom NF2 was a derivative of the single-seat Venom fighter-bomber, the FB1 and, like its forebear, was powered by a single engine. After the NF Meteor, the chief differences from a pilot's viewpoint were the penalty in performance – from having only one engine – and the side-by-side seating arrangement: the latter a welcome improvement in most respects, although making for a cramped cockpit. As for the single engine in terms of added risk; if you've grown accustomed to a second engine, you're bound to miss it to begin with. But more of that in a moment.

At that stage, and quite understandably, the unit lacked the everyday pace of a fully operational squadron. The winter weather, chiefly in terms of early morning fog, did not help matters either. In my first month I flew only nine sorties.

But, to some extent, what that month lacked in quantity, it made up for in character. On my third sortie, the Ghost engine decided to live up to its name. It died on me. We were climbing through 25,000 feet, when, after a few seconds of stop-go surging, it simply wound down and there was an eerie silence. My Nav/Rad and I glanced at one another. I could barely make out his features behind his tinted visor, but the look he gave me seemed to ask 'what now?' Indeed, if that was his question, it echoed my feelings precisely. After all, it was the only engine we had; we were some distance out over a cold North Sea; and in those days immersion suits had barely been heard of. I called Driffield to tell them of our problem and to say that I would be attempting to relight the engine once we had lost some height.

To our great relief, the engine relit at the first attempt. For the technically minded, this had entailed putting the fuel pump on to full stroke, which meant that our idling revs as we crossed the boundary fence on landing were quite a bit higher than normal. No wonder the speed was slow to drop as I held off, and that we touched down so late. But, with some energetic braking, I managed to avoid running off the end of the runway and the indignity which that would have added to the afternoon's events. The fact that it cost the RAF a new set of brake pads seemed to me relatively small beer. The cause of the engine failure – an unconnected wire at the back of a small switch.

In December, having been invited back to West Malling for my official farewell party, I was given the use of the station's Anson in which to make the trip. The understanding being that I would take a whole crowd of people eager to go south for the weekend. I had not flown an Anson since my Wings training in 1943, but felt that by studying the Pilot's Notes beforehand and taking things steadily, that should present no obstacle.

What I had not bargained for was the weather. Driffield was enveloped in fog. To the extent that, when I called for permission to taxi, I was genuinely surprised to receive the go ahead. In starting up, I had really only been showing willing, since I knew how disappointed my passengers were going to be when the trip was called off. The thought of lifting off straight on to instruments in an aircraft I had last flown twelve years before was hardly taking things steadily. But I had asked to taxi and having received permission, I could hardly demur in the next breath.

Picking up just enough of the taxi-way centre line to keep going, we gingerly made our way out to the runway, and proceeded with the engine checks. Again, these provided no reason to call things off, and when I asked for and got permission to take-off, I knew the chips were down. With the engines at full power, I released the brakes and we lurched forward into the gloom. Fortunately, as when taxiing out, I was able to pick up a strip at a time of the runway centre line by which to keep straight. And as I lifted off, and we were engulfed in the 'clag', I locked on to the flight instruments. Never have a pair of eyes flitted more frequently between the directional gyro, the artificial horizon, the airspeed indicator and the vertical speed indicator. In the event, with everything happening in slow motion, as compared with the jet, the ten minutes we spent in cloud gave me no problems whatsoever. The challenge had greatly exceeded the reality. But I don't mind admitting, when we did emerge into the sunlight, I felt a modest sense of achievement.

I did not remain with 219 Squadron for long. At the end of February 1956 my medical category was downgraded and I was removed from flying duties for an unspecified period. It was a blow. But, with the prospect of regaining my aircrew category in due course, I got on with the readjustment. My squadron and station commanders could not have been more helpful. Thanks largely to the latter, Group Captain Bill Moseby, I remained in fighter command. Not only that – I was given command of a small, but important, ground station.

11   *Radar Station*

RAF St Margaret's Bay
In December 1956, after completing a fighter controller's course and doing a stint as a controller at a GCI station, I took command of No 1 Signals Unit at RAF St Margaret's Bay: a twenty-four-hour radar-reporting station on the Kent coast, four miles north of Dover.

By then, the stately, fixed arrays of the 'Chain Home' radar system, which had played such a vital part in the Battle of Britain, were no more than relics of the past in the context of air defence. And although the towers themselves were still in position, fulfilling a GPO communications function, a new operational site had been established close to the Dover Patrol memorial.

Our job was to pick up targets at long range and, in that sense, we were a front-line element of the country's air defence system. Under my command I had upwards of two hundred persons; enough to maintain our operational function on a continuous basis. Although there were no airwomen on strength, four of my officers were WAAF – each of them heading up one of the operational shifts. The majority of my airmen were National Service conscripts, and by and large, I could hardly have wished for a better motivated and well-behaved bunch.

My personal living quarters – which went with the job – comprised a two-storey house perched on top of the white cliffs immediately alongside the operational site, and within a hundred yards of the nearest radar aerial. It was a position fully exposed to the elements, and when the wind had any strength, I could hear it struggling with the huge rotating array of my Type-80 aerial as I lay in bed at night. No doubt that is why the original owner of the house had sold out to the Air Ministry. Not that that whistle-cum-stifled-shriek ever kept me awake: rather the reverse I might say. On windy nights, that sound told me that we were 'on the air' and that all was well. If I did stir in my sleep, it was usually because the Type-80 had temporarily *stopped* rotating.

Other features of my cliff-top environment were the powerful beam from the South Foreland lighthouse – which swept my bedroom walls throughout the night – and the deep boom of the South Goodwins foghorn which often accompanied it. These, too, I rapidly embraced as friendly background.

As a vantage point from which to view the shipping passing through the Dover Strait, the house could hardly be bettered. From my sitting room, the average count was in the order of eight vessels, excluding the South Goodwins light vessel permanently anchored off-shore. And on those days when we were blessed with really good visibility, not only did that number more than double, but with my binoculars, I could also make out the clock tower in the port of Calais, over twenty miles away.

Life deep underground in the operations room conveyed a different atmosphere. Racks of equipment in efficient-looking grey cabinets flanked the walls; shining linoleum covered the floors; and in pride of place was the huge plotting table – illuminated from below and the focus of attention, as the plotters responded to instructions from the officer in charge of the watch. In essence, the job of the reporting team was to look for the *unusual* radar response; something outside the normal pattern of civil and friendly military traffic; something unnotified – flying fast and high, for instance. In a separate room we had the means of controlling aircraft: a small commitment compared with our reporting function.

Each type of radar aerial is designed for a specific job – be it height finding, the control of fighters or, as with the Type-80, picking up targets at long range. But sometimes an aerial yields a bonus. In the case of our Type-80, that bonus was its ability to pick up shipping movements in the Strait when the array was tilted downwards. This was not our proper function, and normally we did not engage in such novel pursuits. But occasionally, at the special request of the Coastguard Service, we did use the Type-80 in this way – when there had been a collision in the Strait in conditions of dense fog for example – and once or twice we did so to solid effect.

The Dover Strait was also a source of recreation, as I discovered after meeting the owner of *José*, a small ocean-going yacht. The said owner was Bobby Melhuish, the local doctor, who stood a shade under six feet six, with long arms to match and hands the size of small shovels: useful attributes when hauling on sheets (ropes) and putting a reef in the mainsail – or, come to that, any number of things required when controlling a sailing boat in a rising wind.

Despite years of sailing experience – including participation in the challenging Fastnet race – there was something endearingly relaxed about Bobby's attitude to sailing. Otherwise I suppose he would not have chosen to sail to Calais with a greenhorn like me as the sole member of his crew. True, we had taken *José* (seven tons Thames Measurement) around the South Goodwins light vessel the weekend before. But it was hardly a real test of me, or maybe the soundness of the boat's equipment, come to that. In fact, as I went to hoist the mainsail on leaving Dover harbour on that first trip, the lifting line had simply parted in my hands: an indication perhaps that things might have been checked out with a little more care beforehand. But to give Bobby his due, as I started to apologise for my 'heavy handedness', he at least interrupted with an apology for taking me to sea with rotten gear.

Regardless of these hints of laxity over matters of detail, my confidence in Bobby's captaincy remained intact. Beneath that casual exterior there just had to be a well of imperturbable calm which, wedded to his experience, would see us through any crisis. So that when, a week later, he broached the idea of sailing to Calais the following morning, I accepted without hesitation.

I awoke the next day to the sound of the foghorn; and to judge from the thick sea mist in St Margaret's Bay at that stage, our trip would have to be postponed. However, neither of us wanted to be the one to call it off, and we decided to take the final decision at Dover harbour, where *José* was moored in the submarine basin.

Miraculously, once aboard, the weather began to clear, and we were soon on our way. An hour later, with a brisk wind behind us and in strong sunshine, we were sailing goose-winged, with the mainsail and the jib filled out like an enormous bra: making a good six to seven knots, and our wake emitting a satisfying gurgle.

Mid-channel, we cleaved our way through a pack of East European fishing vessels, apparently trawling for herring. They were steering a racecourse pattern, queuing up to take their turn like aircraft in an airfield circuit. Steam having to give way to sail or not, they resented our intrusion it seemed: assuming people of those nationalities are not given to waving their greetings with a clenched fist. We gave them the benefit of the doubt, however, and waved cheerily in return. Anyway, with a strong wind at our backs, we were through them in no time.

Having rowed ashore from our mooring in Calais harbour, and secured Bobby's supply of grog and Gauloises (the underlying reason for our visit) we made for the local yacht club: still very much in our sailing rig – half Wellingtons and the rest. From the warmth of our welcome, it was clear that Bobby had been there before. A fact reflected in the scope as well as the duration of the hospitality: a bar dance was soon in progress and half Wellingtons or not we were not allowed to sit out for long. It was well after midnight when we said our farewells. And to put their seal on things, as we left harbour the following morning, the club gave us a rousing send-off with their starting gun.

It was a visit to remember in other ways too: such as our dodgy journey back to *José* after our night out, with her little pram dinghy so weighed down with bottles of rum and red wine that the amount of freeboard was reduced to a few hazardous inches. Bobby merely chuckled at the suggestion that we were in danger of shipping a wave and going down with our precious cargo. As it was the gods were with us, and somehow we got ourselves and our loot safely aboard without as much as a wet sock. But cold sober, I fancy even Bobby might not have attempted such a feat. Another memory of that occasion – since we had tied up in the outer harbour – was

José's nauseous roll the night through: something which, with my hangover, had me draped over the stern well before breakfast.

On departure, a stiff breeze blowing directly into the harbour forced us to make our way out on a series of short tacks, coming almost eyeball to eyeball with a glum-looking row of rod-and-line fishermen lining one of the harbour walls. Unlike the herring fishermen of the day before, they did not shake a fist, but at that range a dirty look can convey a whole lot.

The weather in the Strait was much as it had been the day before. Only this time the wind was working against us, and for something like three-quarters of the trip we had to tack our way across: with Bobby down below attending to the navigation for much of the time. It was now his turn to hang his head over the side, as he suddenly appeared in the cockpit with a cry of 'look out!' To see such an old salt do that cheered me no end, after my pre-breakfast exhibition.

By the time we were within two or three miles of the English coast, the wind had fallen to a light breeze and a sea mist had cut visibility to three hundred yards. At this juncture, we narrowly escaped being run down by the Folkestone ferry, as it suddenly loomed out of the mist like a mobile block of flats. But apart from scaring us, it also told us that we had held our latest westerly tack a bit too long, and that it would probably take another hour to complete the journey if we relied on sail alone. There was also the danger of being rammed the longer we took; and for my money, our tiny masthead radar reflector offered flimsy protection against such a possibility.

With some reluctance, Bobby started the engine. Half an hour later, in the darkness of early evening, Dover harbour acknowledged my torchlight signal that *José* was about to enter. When my batman called me the next morning, I could hardly believe that it was only forty-eight hours since I had motored down to the submarine basin.

During my ground tour I did what I could to remain in touch with flying. Even during my fighter controller's course at Middle Wallop, shortly after leaving 219 Squadron, I managed to steal a little time at the controls of the Balliol target aircraft, thanks to one of my former students. And this 'old boy' approach continued to produce an occasional flight: whether it was in a Meteor T7 with 25 Squadron at West Malling or an occasional flight in a Chipmunk at Manston. But by the time I had completed my assignment at St Margaret's Bay, I had been out of regular flying practice on jet aircraft for just over two years. So in March 1958, with my medical category officially restored, I began a flying refresher course; the venue – RAF Worksop; the aircraft – Meteors T7 and F8.

I took my leave of St Margaret's Bay with a deepened respect for the officers and men of the RAF Control and Reporting branch, and with some regret at being parted from friends in the civilian community. But I did so with a glad heart – after all, pilots belong in cockpits.

12 Examining Duties

CFS Advanced Standards

At the end of my flying refresher course, the long arm of CFS reached out and I found myself back at Little Rissington. On this occasion as the leader of the Advanced Standards team: one of three teams responsible for monitoring flying and flying instructional standards, and for testing Instrument Rating examiners. The other teams dealt with the Basic and Multi-Engine aspects.

My team's examining remit within the UK was to make an annual visit to those RAF and RN units equipped with advanced training aircraft, such as the Meteor T7, the Vampire Mks T11 and T22, and the Hunter Mks T7 and T8. Responsibility for the Canberra T4, although an advanced training aircraft, lay with the Multi Standards team.

Functionally CFS Standards was little more than the former Examining Wing in new clothes: presumably to soften its image. The only real change being that Standards had become part of the Flying Wing and no longer had a separate existence and a Wing Commander of its own. As I write this, the CFS Examining Wing is very much back in business. But politics being politics, who can tell what cosmetic changes the future might hold? What no one in authority has seriously challenged, a far as I am aware, is the usefulness, indeed the necessity, of the examining role in maintaining standards.

A week after my arrival (in mid-May 1958), I was leading my first examining visit. It was to RAF Manby in Lincolnshire. Squadron Leader Bob Eyre took us there in Rissington's Anson Mk 21 and dumped us on the tarmac, leaving us prey to the locals: or vice versa, depending on your point of view.

That first visit was to be quite untypical, in one respect at least. In our wash-up discussion, the aim of which was to discuss our general findings in as helpful a spirit as possible, the Wing Commander Flying was loath to accept even the slightest hint of criticism, thereby denying himself, and us, the opportunity of anything resembling a constructive dialogue. Almost without exception, subsequent wash-ups seemed a doddle after that inhibited climax.

Back at Rissington, and with the Manby visit report in the Out tray, I set about expanding the tools of my trade. First on my list, by way of adding something I had not previously flown, was the Hunter. What a delight it was to get my hands on that aircraft; beautiful in line and sweet to handle; stimulating too, in terms of its shortish duration and the requirement to be on top of its emergency procedures. It also gave me my first experience of powered controls and supersonic flight.

Next came a refresher flight in the Vampire T11: an aircraft I had not flown since I was last at Rissington. Lieutenant Arthur Milnes, the RN member of my team, talked me through the essentials of its flight envelope in a brisk thirty-minute sortie. Such flights also gave me a further insight into my team members' personalities: and vice versa, no doubt. Earlier, I had flown with Flight Lieutenant Derek Hankin in the Meteor T7, and although his temperament and that of Arthur were quite distinctive, one thing they had in common was the high degree of professionalism which went with the job.

In the weeks that followed – although they were outside my examining responsibility as such – I also took the opportunity of flying both the Jet Provost and the Canberra: the other aircraft used by CFS at that time.

In addition to their visits programmes within the UK, the Standards teams made annual visits overseas: some, straightforward working visits to RAF units; others, to liaise with foreign air forces, with the aim of widening our knowledge of how they operated; and a third category, also to foreign air forces, where we were invited to perform the examining function.

During 1958, I made two overseas visits: one to Zweibrücken, leading my team; and the other to various RAF operational units in the Middle East as a member of a combined Standards team. Both visits had their moments.

Arriving at Zweibrücken in West Germany, an RCAF fighter base equipped with single-seat Sabre 6 aircraft, we did so 'cold'. Apart from the air traffic notification of our impending arrival, it appeared that no one at Zweibrücken had the faintest clue who we were or what we were up to. Ostensibly, HQ Flying Training Command – our sponsoring authority – had cleared the visit with the Canadian authorities. Perhaps they had. But, if so, someone along the line had neglected his In tray.

Thank God the average airman is a flexible soul, and that the Base Commander was no exception. He promptly invited us to join a party they were holding in the Mess that night, and the next day people at our own level rescued the working situation with a crew-room chat on tactics, followed the day after by a flight in their single-seat Sabres for every member of my team: an overnight swotting session with the Pilot's Notes and an assurance that we were in current practice on the Hunter was enough to satisfy the authorising officer.

As I settled into the climb after lift-off, I found the Sabre 6 even more twitchy on the aileron controls than the Hunter had been on my first flight. But that was purely my lack of experience on the type, and as the flight progressed that twitchiness gradually disappeared. That apart, the Sabre handled smoothly and positively, with everything in the cockpit coming nicely to hand. It seemed very much a pilots' aircraft, and one could sense why it enjoyed such a handsome reputation.

For the Middle East visit, I settled myself into the co-pilot's seat of a Canberra T4 alongside Squadron Leader John Horwood, the Multi Standards leader. The month was November. A good time to be quitting foggy England, and one where the days in Aden had had time to lose their stifling, furnace-like temperatures of summer. Our first working stop was Luqa in Malta which, with the long legs of our Canberra, we made in a direct flight. The remainder of the visit followed the route – Akrotiri (Cyprus) – El Adam (Libya) – Khartoum – Khormaksar (Aden) and return: with working stops at Akrotiri, Nicosia and Khormaksar.

Outbound, nothing dramatic happened. At the working stops there were plenty of tests to be done, and things of interest to take in by way of local operating techniques. Practically without exception, the people we tested and their superiors took an adult view of the CFS role and got the best out of our visit. In Malta, the testing over, our hosts insisted on taking us to the 'Gut', where we downed a few beers and received the usual saucy remarks from 'Sparrow' and the rest of the girls. In Cyprus, we risked an evening visit to the Key Club in Nicosia, escorted by our hosts and with revolvers strapped to our waists, just in case we met up with EOKA. We met old friends, and made new ones. And we saw some exotic sights: including hundreds of flamingos rising from a lake in Cyprus like a great pink cloud. The drama came on the homeward leg, about twenty-five minutes out of Luqa.

I was at the controls at the time, skirting the anvil tops of some cumulo-nimbus (thunderstorm) clouds, flying at forty thousand feet or thereabouts. As we flew through the innocent-looking, translucent wisps surrounding one of these towering cu nimbs, we suddenly encountered the most violent turbulence. The whole aircraft shook as though gripped by a giant fist, and the needle of the accelerometer fluctuated wildly between a positive and negative reading. Within ten to fifteen seconds both engines had flamed out. The shaking continued for another fifteen seconds or so and increased to such an intensity that if a wing had come off, I would not have been at all surprised. I really thought our time had come, and half expected to hear the navigator's ejection seat fire at any moment. And to judge from John Horwood's initial exclamations of 'Christ! Christ!' as he took control, I gathered he was a bit perturbed as well.

Meantime, Les Stapleton, our navigator, far form abandoning us (neither John nor I had an ejection seat), had been quick to size up the situation. Recognising Bizerta as the nearest suitable airfield, he gave John a heading to steer, and with John's instant agreement, put out a distress call informing Bizerta of our problem and that we were heading in their direction.

With the situation we faced, it was no time for niceties such as losing height before attempting to restart our engines. In fact John had instructed me to do so whilst we were still being hammered by the cu nimb. In spite of our altitude – probably 35,000 feet plus – one engine responded quite quickly and without incident. The other one, however, accumulated a lot of fuel before doing so, with the result that the jet pipe temperature soared to the stops and remained there for some anxious moments before it settled down within the normal range.

We landed at Bizerta, a French Air Force base on the Tunisian coast, without further problems. And having waited for the engines to cool and persuaded a tall, thin mechanic to climb up the jet pipe of the overheated engine, we stood by for his verdict. It was a nail-biting minute or two. Much as I was still enjoying the warm glow of being safely on the deck after the attentions of the cu nimb, I had no wish to be stuck at Bizerta waiting for a replacement engine. Finally the mechanic wriggled himself free and reported no wrinkling of the jet pipe or other signs of damage. Meantime we had peered, as best we could, at the shroud encasing the engine and could see nothing untoward. That was good enough for John. After topping up our fuel tanks, we were on our way; making Lyneham that evening.

In fact, there had been some damage to the engine shroud: as a deeper examination revealed. But given the circumstances, no blame was attached to John Horwood. Once more, those in authority at CFS – in this case, Group Captain George Petty, the Chief Instructor – had stood squarely behind their staff when the chips were down. As to the cause of our double-engine failure: having noted the outside air temperature at the time, and found it to be above the critical figure at which shock waves might be expected to stall the compressor, we concluded that the answer lay in the sheer violence of the turbulence.

If I had to pick one or two personal highlights from my team's

ABOVE. 25 Squadron Meteor NF12. *(Photo courtesy of Harry Holmes).*
BELOW. 85 Squadron Meteor NF14 *(Photo courtesy of Avro International – Ian Lowe).*

RIGHT. José off South Goodwin, 1956.
BELOW. CFS Vampire T11 *Photo
courtesy of Harry Holmes).*

ABOVE. Hunter F1 *(Photo courtesy of Avro International – Ian Lowe).*
BELOW. Specially equipped Hunter T7 used in inverted spinning trials,
1960-61.

ABOVE. Freedom of Louth ceremony, 1965. The Mayor escorted by the author. BELOW. **RAF Manby School of Refresher Flying** Jet Provost T4 flight line. *(See also pictures facing page 266)*

1958 visits programme within the UK, I suppose I would have to select my first flights in the Seahawk and the Javelin.

The Seahawk came my way courtesy of friends in dark blue at Lossiemouth. Being a single-seater, preparations comprised a competent briefing and some over-the-shoulder advice during start up: with the Pilot's Notes slipped into my overalls for possible consultation in flight. The ride would not have been more pleasant and straightforward, and those notes stayed in my pocket. I spent an hour putting the Seahawk through its paces, and not once did it bridle: certainly no aircraft could have entered and recovered from a practice spin in a more orthodox fashion.

By comparison, my familiarisation flight in the Javelin seemed a little tame. I was glad of the opportunity, nevertheless, to climb aboard the eventual successor to our 25 Squadron Meteors. My abiding memory of that Javelin flight – the sheer size of the aircraft apart – was the comfortable way it settled down on the Leuchars GCA glide path, as if it had been designed with that one thing in mind: a characteristic which would have been a great comfort on a dark night or even in daylight with a low cloud base. It takes much more than that of course to make a successful all-weather fighter, but predictability in the home straight is a reassuring feature when you're still at forty thousand feet deciding whether or not you have enough fuel for another interception.

In January 1959, at the Chief Instructor's request, I visited Hawker's airfield at Dunsfold, where the company's test pilots were engaged in spinning trials on the Hunter. At that stage, I suspect Hawkers were conducting these on a private venture basis. Their object being, in the wake of several incidents, to reassure the RAF of the aircraft's ability to recover from inadvertent spins. As the year progressed, RAF involvement increased. But I will let the story unfold a stage at a time.

On that first visit to Dunsfold, after a preliminary discussion, the deputy chief test pilot, Hugh Merewether, and I climbed aboard the trainer version of the Hunter, the T7, for a spinning demonstration sortie. With me following through on the controls, Hugh demonstrated an erect spin and recovery, and then talked me through one in the opposite direction. Using the normal controls deflections – stick hard back, and full rudder – plus the assistance of full out-spin aileron, the Hunter made a nice clean entry. Recovery, too was straightforward: although, following Hugh's advice, I paid special attention to centralising the aileron control as I eased the stick forward. I was impressed. Hugh then invited me to have a go on my own. Since inadvertent spins normally occur during manoeuvre, I said I proposed to enter the spin off a deliberately mishandled stall turn: something we often did in the Meteor T7 to add realism to the spinning exercise. Hugh agreed.

With the aircraft thirty degrees or so off the vertical, I waited until we had almost run out of speed and then, as I pulled back on the stick to achieve a vertical position, I applied full rudder: a classic combination for achieving an erect spin – aircraft close to the point of stall, stick hard back and full deflection on the rudder. But this was no Meteor. The Hunter merely flopped onto its back, and for what seemed an age it just hung there whilst I waited for it to fall through to the vertical; by now I had abandoned the attempt to enter a spin. This behaviour of hanging upside down in an apparently stable situation was new in my experience – slightly eerie too – and, rather than just sit it out, I gingerly applied aileron to see if I could roll the right way up. That proved to be a mistake. At our low speed, it came as no real surprise that the ailerons should not be effective in a roll. But what did surprise me was the strength of the yaw which this immediately induced. Still inverted, the aircraft began to rotate. Hugh asked, 'You know what this is, don't you?' I said, 'Yes, an inverted spin.' With that, he came on the controls, and from there on I followed his control movements.

The events of the next fifty seconds or so – all the time it took us to descend from 36,000 to 18,000 feet – are engraved on my memory. Hugh handled the situation with laboratory-like calm, announcing every move he was making as he set about the recovery. First he pointed to the turn needle, which was displaced hard to one side, and applied full rudder in the opposite direction. It was agreed that whatever else he might try, we would continue to apply full rudder in that direction. Next he dealt with the position of the stick: which, initially, he held back. Then he tried pro-spin aileron. And so it went on: with the spin becoming well developed and once or twice producing a brief but quite vicious period of rotation – something which, in our inverted position, rammed the blood uncomfortably into our heads.

In all, we must have gone through ten to twelve rotations before Hugh was able to check the auto-rotation. From there it was a simple matter of recovering from a steep dive. In retrospect, neither of us could confidently state the exact position of the stick which finally triggered the spin recovery: because, throughout the time Hugh was maintaining full opposite rudder and experimenting with pro-spin aileron, he was also easing the stick progressively forward. Another factor which might have played a part in the timing of our recovery was the fact that we were descending into thicker air, where the controls could be expected to be more effective.

We had not, of course, set out to investigate inverted spinning. Nevertheless, the fact that Hugh's recovery method had worked, at least provided the basis for improving the existing recommended recovery action. Chiefly, it suggested that the stick should be forward rather than back: the use of pro-spin aileron having already been advocated.

Once out of the spin – and with both of us feeling a bit groggy – we selected oxygen to high flow and headed for Dunsfold: we had had quite enough spinning for one day. If my memory serves me correctly, the amount of negative 'g' recorded on the accelerometer was three-and-a-half: its effects being sufficient to give me a distinct headache as the day progressed.

I returned to Rissington favourably impressed with the Hunter's erect spinning and recovery characteristics, but with the realisation that if the RAF was to practise Hunter spinning, it would need to be handled with special stringency. In particular, I was convinced that entry off manoeuvre would need to be ruled out, at least until recovery from the inverted spin had been more thoroughly investigated. The events of that day had also made me aware of the special significance of the Hunter's ailerons, and their capacity to induce yaw under certain conditions.

Over the following months, CFS involvement with Hunter spinning strengthened. In February I revisited Dunsfold – on this occasion accompanied by Flight Lieutenant John Hardaker, one of my Standards team – and in March I attended a meeting on the subject hosted by the Handling squadron at the Aircraft and Armament Experimental Establishment (A and AEE) RAF Boscombe Down. The decision to indoctrinate service pilots in Hunter (erect) spin recovery on a broader scale was taken at Air Ministry level some time before the end of July.

Meantime, Squadron Leaders Crowley and Harrison of the Empire Test Pilots' School (ETPS) had been thoroughly indoctrinated in erect spin recovery by Hawkers, with whom they had also evolved a demonstration sequence for use in the Hunter T7: this sequence included recoveries from the stall, as a lead-in to recovery from the developed spin. Entry to the latter was initiated essentially from a straight and level altitude, with entry from manoeuvre being studiously avoided.

After brushing up our handling on the CFS Hunter F4 at RAF Kemble, John Hardaker and I spent two whole days at the ETPS: both of us going through that demonstration sequence in the air on five separate occasions with either Crowley or Harrison in the Hunter T7, and also satisfying ourselves as to its soundness from an instructional viewpoint.

A week later, on 26th August, John and I began the indoctrination of the staff at the Fighter Command Hunter OCU at RAF Chivenor. Before lunchtime that day John was dead, having failed to survive an ejection from a Hunter T7 after inadvertently entering an inverted spin. I received the news with some incredulity as well as great sadness. Apart from his natural skill, John was a meticulous pilot, and I could not believe that he would have conducted the instructional sequence other than as we had agreed, and which until then had proved entirely successful. But fate is a fickle thing. By the time John was ejecting on his first sortie, I had landed without incident from my second. The immediate inference was that the Hunter's entry into a deliberately initiated erect spin was perhaps less predictable than we and our colleagues at ETPS had concluded and a decision was taken to suspend the indoctrination forthwith.

In September of the following year, comprehensive inverted spinning trials began on the Hunter T7. These were conducted by Hawker Aircraft Limited with ETPS participation, Hugh Merewether acting as first pilot throughout. These trials were spread over a period of twelve months, and included inverted spinning demonstration sorties with pilots from A & AEE and CFS. In terms of recovery action, the trials outcome showed a complete identity between the erect and the inverted spin – full rudder to oppose the yaw, stick fully forward, and ailerons neutral.

Over a six-week period from late September 1959, I led a combined Standards team to the Far East. Our task – to visit RAF units at Singapore and, on the return journey, also to make working visits to the Ceylonese and Pakistani air forces. We took our own transport, the trusty Varsity, with Flight Lieutenant Les Stapleton at the controls.

There was no question of arriving in Singapore with jet lag and needing a day or two to recover before getting on with the job. With six overnight stops, plus a day's crew break en route at Karachi, we passengers at least, stepped off at Changi not only with our marbles properly in place but eager to make a start. By the airline standards of today, involving perhaps only a single stop en route, our journey by Varsity is made to look like the old-fashioned stopping train: Little Rissington – Orange – Luqa – Nicosia – Diyarbakir – Teheran – Sharjah – Karachi – Delhi – Calcutta – Rangoon – Bangkok – Changi.

On this occasion, there had been no question of tangling en route with cu nimb clouds at forty thousand feet. Our cruising height was a much more modest affair, from where we threaded our way through the weather with the assistance of our on-board radar. I suppose our afternoon leg from Rangoon to Bangkok gave us as much turbulence as anywhere, when we received some buffeting from the dying remnants of the Inter-Tropical Front (ITF) as we crossed a ridge of the Thai hills. But it was nothing more than a gentle reminder that the local weather had still to be reckoned with.

We were two weeks in the clammy heat of Singapore: where few days pass without a rain storm in the late afternoon. And when it rains in Singapore, stair rods are not in it. It teems down for minutes on end; soaking everything and briefly filling the storm drains. I think it must be the only place where I've seen RAF personnel in uniform carrying umbrellas. Whatever else the RAF may be guilty of, it is not lacking in adaptability.

On reporting to HQ Far East Air Force (FEAF), I again met up with John Barraclough. This time in his capacity as Group Captain Operations. I especially recall him seeking the opinion of the Multi members of my team on the local operating criteria for the twin-engined Pioneer: a matter they looked at in practical terms during their working visits up country. My own immediate contribution being to carry out a number of pilot standardisation flights in the Meteor T7.

With Christmas threatening, I suppose there was not a single member of our team who did not spend an evening or two ogling the cameras and Noritaki dinner services on sale in Changi village. After UK prices, it seemed almost wicked not to seize the opportunity of such purchases. But as one who had fought the Japanese, it struck me as ironic that it should be the goods of the defeated nation that represented the bargains. The fact remains that by the time we left for the Nicobar Islands, en route for Ceylon, the empty spaces in the Varsity were rapidly filling with goods from Japan.

Before landing at Katunayake, I had made only fleeting visits to Ceylon and had met few of the islanders themselves. But by the time we left a week later, the members of the Royal Ceylon Air Force (RCeyAF) had completely won us over with their friendliness and courteous manners. A very young air force, still finding its feet, it was commanded at that stage by Air Vice Marshal Barker, an RAF officer on secondment. He seemed made for the job; vigorous and forceful in speaking up for his progeny, but considerate and encouraging in the way he handled them.

Our hosts were equipped with four aircraft types – Chipmunk, Balliol, Heron and the single-engined Pioneer. This gave me no personal involvement in the testing and recategorisation flights as such, and our hosts, not wishing to see me left out of the flying programme, encouraged me to sample their hardware as the opportunity arose. The Heron was too committed for that to come my way, but I flew each of the other aircraft on offer.

Part of the penalty CFS pilots pay is that rather a lot is expected of them. In reply to my request to rejoin the Katunayake circuit on my first solo flight in the Balliol, the controller cheerfully asked me if I would put on an aerobatic display before I landed. What a request! Strange aircraft; expectant audience; loss of face for CFS if I refused; big clanger if I accepted and anything went wrong. 'Yes,' I said, after a moment's hesitation, 'certainly, but please don't expect too much.' Fifteen minutes later, I was landing to warm applause. Maybe some of their enthusiasm for my off-the-cuff effort stemmed from sheer politeness; I don't know. But how glad I was to have a well honed Harvard sequence to call on.

I hasten to add that whilst I was filling in time with this scalp hunting, the rest of my team were hard at work on the real business of our visit: recategorisation, standardisation and instrument-rating renewals. At the end of quite a busy week, and with the wash-up conference behind us, we crossed to Trincomalee in the Heron; there to be hosted by the RCeyAF commander and to swim within the safety of the shark net which fenced off his seawater pool from the remainder of the Indian Ocean. Such moments of recreation worked wonders in recharging the team's batteries.

Arriving at Mauripur (Karachi), we left our Varsity in the care of the Pakistan Air Force (PAF) and boarded one of their Bristol Wayfarers for the PAF College at Risalpur on the North West frontier.

En route, we called at Quetta to drop off a commodore of the Pakistan Navy. It struck me, as he stepped briskly away in his whites, that he was not only four hundred miles from the sea but also in rather mountainous country for one of his calling. Since he was carrying a briefcase, presumably he had swopped the bridge for the conference table and the lecture podium.

Leaving Quetta, our Wayfarer continued to drone its way over rugged, often moonlike, country; skirting the Afghanistan border and crossing the Tribal Areas before reaching our destination.

The PAF College, which combined officer and flying training, positively buzzed with keeness and efficiency. A flying programme had already been drafted and required little more than our agreement. The next morning was were quickly airborne and at work.

Unlike our visit to the Royal Ceylon Air Force, I was able to play a full working part in the day-to-day programme. Knowing that the college was equipped with the North American PT6 – their equivalent of our Harvard – I had come prepared. Before leaving the UK, along with Lieutenant White RN of Basic Standards, I had spent two days polishing up my handling on a Harvard 2B; borrowed, as I recall, from RAF Boscombe Down. Sitting in the front seat on those recategorisation flights at Risalpur brought a pleasant whiff of earlier days in the Harvard.

Strolling around the college grounds one evening, what struck me most was the generous scope of its sporting and recreational facilities: including stabling for a dozen or more fine-looking horses. What a superb environment it provided for the cadets. And as an investment in the PAF's future, it seemed to me that it could hardly fail to pay a handsome dividend.

We were now in the last week of October, and in the gin-clear skies above Risalpur, visibility was practically unlimited. A mere sixty miles to the West, ringed by handsome mountains, lay the Khyber Pass. Hardly a sortie went by without my casting a curious eye in that direction. Near the end of our visit, I suggested to one of my charges that we might complete our sortie by making a small diversion to look at the Pass. He politely but firmly turned down the idea, referring vaguely to earlier tensions in the area and the need not to stir things up. Disappointed, I left it at that.

In three days we had completed the tests, and on the fourth morning we held our wash-up discussion. The PAF instructors had put on a very competent performance; fully justifying the expectations of their seniors. The following day a PAF Wayfarer returned us to Maripur.

Four days later, those of my team with Noritake dinner services were explaining to the customs at Lyneham just what was in those big cardboard containers. The principal item among my own loot was a Japanese cine camera, on which I had shot many yards of film since leaving Singapore. But when that film was processed – although it was not a total disaster – I could see that my camera's built-in light meter had been little more use than a child's dummy – and that the manufacturer was no superman after all. I wasn't sure whether to burst into tears or let out a cheer.

November to January are dog days for service flying in the UK, and the winter of 1959–60 was no exception. In the Vale of York, for instance, airfields can be fogbound for days on end, and our strategy was to plan our work for that period around visits to coastal airfields like Lossiemouth and Valley, and otherwise to catch up on in-house work at Rissington.

In February, I joined a team visit to training establishments of the German Air Force. We were led by our chief instructor, Group Captain 'Jimmy' James, a stocky, cheerful man, with an infectious enthusiasm, and a gift for getting the best out of people. We were away for a week, visiting Fustenfeldbruck, Landsberg and Diepholz; our purpose being to exchange ideas on training methods and to gauge the success of one another's systems.

Behind that engaging grin of Jimmy's was a shrewd, analytical mind. It was an education to watch him gently ferret out information on which to base our assessment. One of the yardsticks in measuring the success of flying training methods is the accident rate. And whilst the GAF appeared to be quite open about the figures within their equivalent of our Flying Training Command, it was interesting to note how cagey they were regarding accidents in their front-line squadrons. Indeed, they probably had quite a bit to hide, considering the number of known fatalities among their F104 squadrons at that time.

At the practical level we sampled the Piaggio P149D, their basic trainer, and their Fouga Magister CM 170R interim (jet-powered) trainer: the latter being equipped with an intriguing periscope device to overcome the restricted view from the rear seat.

Socially, our sternest test was to know when to stop accepting the

Steinhagers from the queue of cadets intent on toasting us at one of the evening receptions. A head for drink isn't a *must* in order to succeed in the RAF but, however unfairly, it is sometimes worth a few Brownie points.

March found my team making its annual visit to the RN air station at Yeovilton to carry out standardisation tests in the Navy's Vampire T22 and Hunter T8 aircraft. During our previous visit, I had been asked for my view on the adaptation of the aircraft carrier 'Meat Ball' angle-of-approach indicator to airfield use. 'Wings' (the commander in charge of flying) pointed towards a Venom FAW 21, saying, 'It's all yours. Have a go and let me know what you think.' What a marvellous job I had, being given a new toy to play with, just like that.

I've forgotten just how the Meat Ball was engineered, but to stay on the correct angle of approach one had to keep a ball of light between two horizontal markers. This contraption was placed along-side the runway, some distance in from the threshold, and in effect did the same job as the standard airfield angle-of-approach indicators are designed to do at night. The chief difference was that the Meat Ball stood out more prominently, even in strong sunlight, as well as appearing more sensitive to deviations from the correct flight path.

Two approaches, and I was convinced of its effectiveness. My main concern during the sortie was the damage I might be doing to the environment as I jettisoned my wing tank fuel before calling for a controlled descent. The fuel system had been filled to the gills, and Wings had suggested the jettisoning as an alternative to spending time burning it off. Dumping it from eighteen thousand feet or so, Wings reckoned it would be well and truly dissipated long before it reached deck level. I do hope he was right and that the Greens will not catch up with me for this.

In mid-June of 1960, I was posted to the Air Intelligence branch of the Air Ministry to head up a section – A13(i) – concerned with the assessment and dissemination of intelligence on the air defences of the Warsaw Pact. Apart from the interesting nature of the work, there was cause too for celebration since the job carried with it the

acting rank of Wing Commander: in the event, substantive promotion followed a fortnight later on the half-yearly list.

Leaving CFS, I did so with a sense of fulfilment. The job had
been wonderfully varied and challenging. Excluding derivatives, I
had flown nineteen different aircraft types and visited seven countries: eighteen if one includes those we visited in transit. And I had
been privileged to lead a highly professional and dedicated team.
The one blight on all this, of course, was the sad death of John
Hardaker: a true professional and a man we could ill afford to lose.

In the following New Year's Honours List I was awarded the Air
Force Cross. I mention this for two reasons: first, because I think it
would be coy not to do so – it has after some relevance to my story
– and also because of the intense pleasure it was to give my mother
in attending the ceremony: something I found particularly gratifying,
especially given her initial shyness about doing so. The fact that the
Queen herself performed the investiture on that occasion was, of
course, doubly pleasing to us.

My reference in the penultimate paragraph to the wide variety of
aircraft types I had been privileged to fly whilst at CFS is an excuse
to record the entire list of aircraft I flew during my service career:
these can be found in Appendix B. I realise that this may be seen
as boastful, and I suppose it can be construed that way. Personally
I like to see it as a record of something achieved; something I can
view with satisfaction as I recall the enjoyment it gave me.

13   *Intelligence Staff*

Air Ministry Intelligence

In 1960, our nuclear deterrent was vested in the RAF 'V' Force; Vulcan and Victor bomber aircraft, which could be kept at readiness alongside the runway – as with fighter aircraft – and which could be airborne within minutes of the order to 'scramble'. (To provide an effective deterrent their scramble time had to lie within the flight time of Soviet medium-range ballistic missiles (MRBMs.) In war, their job would be to penetrate the Soviet air defences and to deliver their weapons on preselected targets: a difficult task, but feasible if the crews knew enough of the location and capability of the enemy's radar stations, his surface-to-air missile (SAM) sites, and his interceptor fighter aircraft. The job of my section was to provide and continually update that information.

Initially, I shared an office with Wing Commander Charles Owen, a wartime bomber and pathfinder pilot of distinction who, post war, had also commanded the first 'V' bomber squadron. His job was the mirror image of my own; he and his staff provided Fighter Command with details of the Soviet offensive capability – that of the Russian long range air force and Soviet MRBMs. Thus it was that we were known to our friends as the 'offensive' and 'defensive' wing commanders: Charles rather enjoying the macho connotation of his handle. Later, Wing Commander Mike Stanton assumed the offensive mantle; although anyone of a more gentlemanly bearing it would be hard to imagine.

The free movement of Westerners within Soviet bloc countries being severely restricted at that time made the job of intelligence collection by that means a difficult and often unpleasant task; with patchy results. And in maintaining the enemy Air Order of Battle, we relied to an extent on 'electronic' means.

Today, thanks to developments in surveillance from space, the nature and disposition of air defences – and offensive capabilities too – is largely an open book, with or without the disintegration and opening up of the former Soviet Bloc. But in 1960 there was no such technology. There was indeed no easy way of peeping over the fence. An unscheduled approach to Soviet airspace, let alone an unauthorised overflight, was sufficient to trigger a very hostile reaction. One has only to recall the shooting down of an American Lockheed RB47 aircraft in international airspace over the Barents Sea in 1960 for evidence of that. The introduction of a sophisticated Soviet surface-to-air-missile (SAM) capability around that time (the SAM 2) also made overflights by the high-flying American Lockheed U2 reconnaissance aircraft a questionable risk after Gary Powers was shot down. All in all, it was a time when the Cold War was near to its iciest and intelligence had to be grafted for.

Perhaps the most dramatic piece of intelligence I can recall showed up during the 1962 Cuban missile crisis, when the US government made available positive proof of (Russian provided) offensive missile sites on Cuban soil. This was contained in some excellent U2 photography which Charles Owen's section was asked to verify for Downing Street's benefit. Looking at those photographs hardly made for a good night's sleep given the strident confrontation then taking place between Nikita Khrushchev and John Kennedy: the one hotheaded; the other youthful and with something to prove.

Both Charles Owen and I answered to a deputy director of intelligence (DDI 3), Group Captain RJ 'Beetle' Oxley when I first arrived. But within weeks Group Captain John Aiken took over, and it was under him that I served most of my time in the department. John was an assertive leader who was good at meeting deadlines and persuading his staff to do likewise. It was clear, even then, that lack of drive would not be the quality to hold him back from much higher things. In due course he not only became C-in-C Near East Air Force, but went on to take his seat on the Air Force Board as the member for personnel (AMP). I enjoy working for someone who knows what he wants; tells me what it is; and lets me get on with it. That was John Aiken to a T.

Whilst our routine function was to feed information to the operational commands and to prepare briefs for use within Whitehall departments, occasionally Charles and I were called on to brief top members of the defence and intelligence communities eyeball to eyeball: a job which threw up more than a handful of interesting characters.

Among these were Earl Mountbatten (then CDS) and Sir Solly Zuckerman (Chief Scientific Adviser to the Secretary of State for Defence) whom we briefed jointly on one occasion. Far from being an intimidating experience, the sharp-witted interplay between them made the occasion quite entertaining. These briefings, which usually went like clockwork, were always a challenge however, because, whilst the average VIP very much wanted the presentation kept in broad outline, there remained the possibility that this time round the celebrity would want to prove his grasp of the subject by pitching in with a question on a point of detail; one on which his staff had very likely rehearsed him in the answer. Such persons were in the minority, but their mere existence was enough to keep us on our toes.

One of the week's highlights was to attend a meeting in the Cabinet Offices to inject our piece into the Foreign Office round-up of events affecting the Soviet Union. The meeting was chaired by 'Nico' Henderson who was then in charge of the Western Desk and who went on to do much grander things: as Sir Nicholas Henderson, becoming one of the most distinguished British ambassadors in the post-war era (Warsaw, Paris, Bonn and Washington) in the latter post, after his official retirement date during the Falklands War. Nico's ability to grasp the essentials of a dozen or so diverse inputs and to get something down on paper there and then in succinct, plain language was remarkable. And it was all done with such good humoured charm.

Occasionally, a piece of intelligence fell into our laps without the least risk on anyone's part, the Red Square parade celebrating the October revolution being a case in point. Often the occasion for the first public airing of a new weapon, it would have our brethren from DDI (Tech) gathered eagerly in front of the TV in the duty officer's room, praying for sight of something in that category. One year, I remember it was the sight of a fighter aircraft on a transporter; but minus its undercarriage and cockpit, and obviously adapted for some other purpose – the question was, what purpose? It turned out to be an air-launched stand-off weapon for carriage under the wings of long-range naval aircraft. But how our technical colleagues scratched their heads when it came into view.

It would be nice to say more about my time in intelligence, but I'm afraid the very nature of the work prevents that. Suffice to say that I found the work fascinating, and that while I felt thoroughly frustrated with the bureaucratic ways of the civil service from time to time, I nevertheless came away with considerable respect for many of the people in it. Certainly, I would not willingly have missed the education of serving in Whitehall and the opportunity of occasionally seeing really able people at work; people who combined natural ability with sufficient knowledge of the Whitehall system to achieve shifts in policy in spite of the labyrinth of committees and the all-pervasive Treasury brake.

14 *Flying Station*

RAF Manby

It was almost four years before I left the Air Ministry to resume normal RAF life. But the break from a chairborne existence was hardly a clean one. The first thing confronting me was a course at the Joint Services Staff College at Latimer. In fact I welcomed the idea, since in spite of my years in Whitehall, I lacked a formal 'staff' qualification at the senior level – a must, I was told, if I was to progress much further. Almost as important was the old-boy network it would provide with all three services.

Latimer was a great experience. We had a veritable galaxy of talent by way of visiting speakers – Mountbatten, Vic Feather, AJP Taylor and General Bill Slim, to name but a few. By way of practical education, we visited HMS *Dolphin*, where we watched spellbound as the submariners practised their escape drills in a huge tank; we visited Germany to see an armoured unit of BAOR (British Army of the Rhine) show off its capabilities; and we visited front-line RAF units. In the academic sphere, we sometimes worked well into the night to complete exercises in syndicate; arguing our respective corners – light blue, khaki and dark blue – with passion, if not always with eloquence. And through it all we became friends. Yes, the staff were top rate and nudged us in the right direction, but what made Latimer was the relationship we students built up with one another, as wariness gave way to understanding and respect.

From Latimer I was posted to command 90 Squadron, a tanker (airborne refuelling) unit equipped with Vickers valiants; former 'V' bombers. Being a mobile, self-contained unit, this was regarded as a plum job. But, sadly, it fell through. Overnight, it seemed, serious fatigue problems were revealed in the Valiants' main spars. Bad enough. But to add to my concern, the Air Secretary's branch seemed almost casual about finding me another squadron. Having by now completed by Meteor and Canberra flying refresher courses, I was attached to the RAF college at Cranwell to act as Administrative Co-ordinator for Unison 65, a Commonwealth Chiefs-of-Staff conference. When I remonstrated over this apparent indifference to my future, I was advised to be patient.

Five months later, in September, 1965, the Air Secretary revealed his hand. I was to command RAF Manby – a flying station – with the acting rank of Group Captain. And, as it turned out, the New Year's promotion list was to make me an honest man. 'All things' etc.

Arriving at Manby, I was pitched in at the deep end: within a month the station was due to receive the Freedom of Louth; there were three new types of aircraft for me to master – Jet Provost Mk4, Varsity and Dominie (HS125); and I had yet to make a start on a detailed report on Unison 65, required in short order by the Commandant at Cranwell. All that in addition to running the station. It was just as well that I had the indefatigable Jimmy Douglas as my Wing Commander in charge of administration.

The Freedom ceremony, which before the event I could have done without – not the sentiment behind it you understand, just the timing – proved to be not only a pleasant occasion but a blessing in disguise. It meant that in short order I knew everyone who was anyone in the local community, and – enhancing official links apart – it triggered some lasting friendships. As to the ceremony itself, it went off without a hitch. And bolstered by the presence of the Queen's Colour Squadron, I think my airmen were something of a revelation to the people of Louth with their smartness and bearing. No less important, Air Marshal Sir Patrick Dunn, our C-in-C, also thought we had put on a good show.

The local community was friendly, hospitable and distinctly pro-RAF – none more so than the Dixon family, who first introduced me to game shooting. With Diana Dixon pointing me in the right direction, I sought the advice of an expert before investing in a secondhand gun. A few practice sessions with Diana's contact to get my eye in, and soon I was receiving formal invitations to join in a day's shooting. Not rough shooting, but properly organised affairs where you took your place at a numbered position, moving progressively along the line as the party went from one location to the next, and with the game being driven towards the guns by a small army of beaters. At the end of the day, one went home with a cock and a hen pheasant, and maybe a brace of woodcock.

We learn best from our mistakes, it's said. Perhaps game shooting and I were no exception. With the contents of The Game Shot's Vade Mecum to guide me, I did my best to observe the code of conduct expected of every 'gun' but, whether or not I deserved it, I recall receiving the most frightful ticking off from the 'gun' alongside me towards the end of one such shoot, as I felled a bird almost at his feet. As he saw it, it was his bird I had taken. Moreover, according to his outburst, it was something I had been doing for the best part of the afternoon. When I apologised to my host for upsetting one of his regular guests, he said, 'Oh, old so and so – take no notice, he's always belly aching.' However, after that little fracas, I did try that bit harder not to poach.

As the station commander, my principal function, apart from being responsible for the administration and discipline at Manby and its satellite airfield at Strubby, was to run the School of Refresher Flying (SRF), where we handled pilots of all ranks from Sergeant Pilot to that of Air Vice-Marshal. A job at individual instructor level which called more for skill in restoring what lay dormant than that of straightforward teaching. A consequence of this maturity in our students was to make for a singularly low accident rate and one which seemed to baffle the statisticians at our command headquarters, who once asked me how I accounted for this aberration. It was almost as though they would have slept easier in their beds had we complied with the statistical norm.

Although I now wore a brass hat and occupied the Station Commander's residence, I was by no means king of the castle. Also housed at Manby was the College of Air Warfare and its Commandant, Air

Commodore John Topham – my AOC to boot. My reader may take it that this arrangement called for diplomacy on both sides. Thankfully, John Topham and his successor 'Cyclops' Brown showed considerable understanding for my position. Both were pilots of distinction (John as a wartime nightfighter ace, and Cyclops first as a Battle of Britain fighter pilot – when he lost that eye – and later as a 'V' bomber station commander) and, being men of spirit, neither had room for pettiness. Both, too, were as shrewd as they come, although in other respects they could hardly have been more dissimilar. John – suave, measured, regal, and with a discerning eye for a balance sheet; Cyclops – spontaneous, matter of fact and determined to don his flying suit on a regular basis.

My responsibilities for the college were mainly administrative; although, with so much of its visiting 'brass' coming and going through my airfield, the college programme was seldom out of mind. My most tangible contribution to college activities, housing and catering apart, was to provide pilots and support facilities for a flight of HS125 Dominie aircraft housed at Strubby and used to provide practical training for specialist courses in navigation; a secondary activity within the college.

The Dominie commitment also gave me the opportunity of an occasional flight overseas, when I occupied the right-hand seat to places such as Bodo to the north (beyond the Arctic Circle) and Malta to the south. It was during a visit to Malta that I also managed to get my cook on board, not to improve the in-flight catering but to allow him to look up his father who was serving there in the RAF. The latter could hardly believe his eyes when his son stepped onto the tarmac on our overnight visit.

RAF Manby first opened in 1938, and during the second World War it housed the Empire Air Armament School responsible for bombing and air gunnery training. In 1950 it moved upmarket with the establishment of the RAF Flying College, and it was the latter which really put Manby on the map. Between 1950 and 1957, its students created five distance/speed records in Lincoln and Canberra aircraft, visiting places as widespread as Cape Town, Ottawa and

Tokyo. Locally, the flying college students were able to practise live bombing and air-to-air firing, using Canberras and Hunters.

But well before I set foot in the place, the Treasury had axed this heady stuff. The courses had taken on an academic character, and the college had been renamed the College of Air Warfare. (When I took command, the only connection between Manby's flying task and the college was the specialist navigator training mentioned on the previous page.)

As to the Air Warfare courses: their aim was to provide our future RAF commanders and their air staffs with a working knowledge of contemporary defence problems, as well as to remove any blind spots in their knowledge of air warfare. But although the courses were essentially academic in nature, through periodic visits to the operational commands, interspersed with incoming visits from such in-the-know people as the RAF Cs-in-C and the Vice Chiefs of the three services, the switch of emphasis from the days of the Flying College was at least seasoned with a healthy dash of realism.

The college staff were forward looking too – part of the syllabus looking at the relevance of space to warfare of the future. To such purpose, indeed, that one might argue the staff's treatment of the subject was almost too successful. Gradually they became committed to courses on space not only for the RAF in general, but also for selected individuals of other government agencies. I recall too a space course being run for the benefit of NATO officers.

15 *Air Staff Planning*

RAF Germany

From Manby I went to Rheindahlen in Germany as Group Captain Air Plans at the RAF headquarters (HQ RAFG).

Put ashore at Ostend in the small hours en route to Rheindahlen – short of sleep, and my blood sugar at a low ebb no doubt – it was hardly the best moment for a first-time switch to driving on the right. In my semi-comatosed condition, I had travelled less than half a mile before meeting the opposing traffic head on. The experience was such a shock, it was that no doubt which saved me from ever repeating the error. Fortunately, the confrontation occurred at traffic lights as I was completing a ninety-degree turn, and the opposing vehicles were stationary, waiting for the lights to change. Thus I was able to back off without injury to either side, bar considerable loss of face on my part: the bemused looks from those drivers staying with me for some time.

Arriving at HQ RAFG in November 1967, I could hardly have chosen a more interesting stage at which to take up post. The outline plans for the introduction of the Harrier, the Jaguar and the Buccaneer, were due for fleshing out so that we would be good and ready for the arrival of the aircraft as such. And my arrival also coincided with the NATO decision to adopt a doctrine of flexible response, whereby we had to plan for the 'hardening' (added protection) of our airfields and communications in order to withstand a three-week period of conventional warfare: thereby giving the politicians additional talking time and an alternative to an automatic nuclear exchange. Evidence that the chilling experience of the Cuban missile crisis had not been wasted.

Pending the arrival of the new aircraft types, our airfields on the Rhine housed the Canberra strike force, and our forward airfield at Gütersloh our Lightning fighter squadrons. Before taking up post, I'd spent most of the months of August to October flying both these types, so that I arrived au fait with our current hardware: a well tried RAF principle of staff officers being properly in touch with problems at the sharp end.

At that time, Air Marshal Sir Denis Spotswood was C-in-C RAF Germany; combining this role with that of Commander 2nd Allied Tactical Air Force (2 ATAF). He was the archetypal airman and C-in-C, dark handlebar moustache, steady penetrating gaze and cool, aloof, manner. And there was little his staff could tell him about aircraft and their operation; be it piloting, navigation or weapons systems. In fact that handlebar moustache was his only concession to the popular image of the fun-loving, back-slapping airman. Behind that handlebar was a true, calculating professional, and it was no surprise when, in due course, Sir Denis went on to become Chief of the Air Staff. (CAS).

Given the C-in-C's 2 ATAF responsibilities, the day-to-day running of RAF Germany devolved largely upon Air Vice-Marshal Tony Heward; a shrewd, poker-faced man, who seldom missed a trick and who kept everybody on their toes; headquarters staff and station commanders alike. Tony Heward too, later became a member of the Air Force Board; in his case in the capacity of Air Member for Supply and Organisation (AMSO). (Towards the end of my tour John Aiken took over as deputy commander on promotion to Air Vice-Marshal and although I only served under him for a brief period it was clear that he would be making his customary dynamic contribution.)

And if our two top commanders were devoted to the job, their wives, Ann and Clare, were but a step behind, in support. Which, given the through-put of visiting 'firemen', was just as well. With Whitehall only an hour or two away, the temptation to drop in and see things at first hand was too much for some MOD staff. And

Whitehall politicians too caught the travel bug. One would hardly dare suggest that the prospect of a bottle of duty free, and some Chanel Number Five for the little lady, influenced their plans to any degree, but no doubt such goodies set the seal on a day out of the office or the House, as the case may be.

My immediate boss was the Senior Air Staff Officer (SASO). In my two and a half years at RAFG I served under three SASOs – Air Commodores Gill, Hazlewood and Mellersh; but mainly under Freddie Hazlewood, an energetic man, every bit as imbued with a love of flying as the day on which he had joined up, and in every respect the right man to inspire the operational units. Freddie hadn't quite the same love for paperwork but, here again, he knew what he wanted, and with his touch for delegation, that hardly mattered. He was appreciative of his staff's efforts, and it was a pleasure to work for him.

Generally speaking, my staff and I were given a broad directive and allowed a free rein in evolving our detailed planning proposals. But now and then politics intruded, and we received prior word as to how the C-in-C expected us to handle a particular situation. Thus, logical thought was sometimes adjusted in the wider interest. I have in mind, for instance, the original decision to base the Harriers at Wildenrath, although their shortage of range screamed out for their forward deployment at Gütersloh. (In part, the reason for that particular ploy appeared to stem from the C-in-C's desire to resist a further closure among his Rhein airfields; having been obliged to sacrifice Geilenkirchen in an earlier Treasury cut.)

Such adaptability is, of course, a necessary quality in a planning staff if they are to stay the course; just as it is when an arbitrary Whitehall decision knocks away the bottom row of bricks supporting one's latest master plan. After one such occurrence, I recall Sir Christopher Foxley-Norris (who succeeded Sir Denis Spotswood) seeing fit to counsel me against taking the upheaval too much to heart; not that he was in any sense encouraging a casual approach to the formulation of the replacement plan. Sir Christopher had been a planner himself; he knew of the frustrations at first hand; but he had also learnt how to accept such changes philosophically.

The revolutionary nature of the Harrier, capable as it is of dispensing with formal airfields, presented a formidable challenge to we planners: how would it be supported from its 'hides' in terms of fuel and weapons? Who would guard it? How would the tasking authority communicate with it; and so on. And over and above the matter of logistics and communications hovered the question of which authority would control it in battle – light blue or khaki.

Much had been made of the Harrier's potential as a tank buster, and everyone knew of its limitations in range. So was it not essentially a battlefield weapon? Our colleagues in HQ BAOR seemed to think so. But no self-respecting airman was going to hand over operational control of any part of his force except for a specific objective and for a strictly limited period of time; the use of air power in penny packets has long since been anathema to RAF commanders. Given this importance of Harrier control, I personally spent more than the odd weekend drafting something which might just pass muster with those in khaki as well as with my own C-in-C. Those early attempts to achieve an agreed position had a rather tortuous passage, as received wisdom came under fire from both sides.

At that stage, with the introduction of the RAFG Harriers still a couple of years away, Hawkers intensified their efforts to interest other 2 ATAF airforces in the Harrier. Naturally, HQ RAFG gave what help it could, without wishing to appear too pushy. As part of this sales offensive a Harrier made a spot landing on the grass alongside the Rheindahlen combined Headquarters, witnessed by the brass from 2 ATAF and Rhine Army. Many of the officers were accompanied by their wives in summery dresses, and as the aircraft made its approach the scene smacked a little of Henley, less the champagne. Seconds later, as the ground cushion of air lifted those dresses and the ladies were momentarily enveloped in a shower of dead leaves, I half wondered (as the RAFG co-ordinator) what thanks I was likely to receive. Mercifully, that bit of drama was over in seconds and almost before the Pegasus had wound down I was again breathing easily.

Later, we gave a presentation to SACEUR'S staff at SHAPE Headquarters on how RAFG intended to operate the Harrier.

SACEUR's top airman (an American) received our presentation politely, but one got the firm impression that since American industry had invested heavily in aircraft which relied on formal airfields, he felt duty bound not to show any real enthusiasm for V/STOL (Vertical/Short Take-off and Landing) aircraft. His main contribution was to highlight the problems of logistics; a fair point, and one which we were still addressing at that time. Since then, of course, the RAF has acquired Chinook helicopters, which would be used to overcome the problem of Harrier resupply in the field.

As part of a forward-looking exercise in Harrier deployment in the field, a number of possible operating hides were selected and a scientific study conducted into the likelihood of their detection. The results were quite encouraging. Alongside this study, and in order to get an operator's view on the sites selected, I took a small team from Hawkers into the field and showed them a couple of typical hides; on the strict understanding, naturally, that (even though those particular sites might never be used in anger) any notes they might take were to make absolutely no reference to their whereabouts.

Bill Bedford, who was in charge of the Hawker team was, as ever, brimming with enthusiasm and energy. He was also busy sketching and taking notes; to the extent that, before we parted, I really thumped home to him and his colleagues that it would be more than our lives were worth if they revealed the locations, even by accident. My point evidently went home. Before they boarded the plane at Cologne, Bill sent me a postcard assuring me that they had gone one better and destroyed all their notes, emphasising his loyalty to the Crown by choosing a postcard emblazoned with the Union Flag.

In July 1969 I applied for early retirement. Not that I felt my promotion prospects had run out or that the RAF had gone to pot. But with another dose of Whitehall staff work almost certain to follow my tour in Germany, and with the enticement of a severance handshake, the grass suddenly looked at whole lot greener in the world outside. I had no job lined up: that was something I proposed to tackle during my resettlement leave. But I was never in doubt that I would find a job to my liking. And the thought of stepping into unknown territory had a certain excitement to it. Had I had family responsibilities to consider, no doubt my attitude would have been less casual.

It was February 1970 before I took my leave of HQ RAF Germany on my way out of the RAF. As well as the wrench of finally tearing up my roots, neither did I depart in very good physical shape. At a splendid 'Tramp' supper given by John Aiken's wife Pam a few weeks earlier, I had broken an Achilles tendon dancing a Scottish reel and was still hobbling about with the use of sticks.

The fact that I was repatriated by air and spared a difficult homeward journey by rail and ship was, I'm sure, ultimately thanks to my C in C. The Senior Air Staff Officer too – Air Commodore 'Togs' Mellersh – helped to brighten my departure from RAF Germany by coming along to Wildenrath airfield in person to see me off, accompanied by 'Paddy' Simpson, my deputy in Air Plans. These gestures were ones I much appreciated, demonstrating as they did that (in spite of its increasingly business-oriented outlook) the RAF still had a good deal of heart.

Within days of returning to the UK, I was at RAF Hedley Court with the wasted muscles of my leg being worked on by the rehabilitation staff. In fact it would be more correct to say that I was doing the work and they were applying the stick and the carrot. They did a magnificent job, but the last thing they would tolerate was any slacking in the ranks; certainly anyone expecting sympathy got short shrift. Even a fellow who'd recently lost a leg received the same brisk treatment. After a week of this regime I (quite genuinely) pleaded a prior commitment and headed off to the RAF Club and a Business Resettlement course.

It was something of a watershed to be parting company with the RAF and twenty-five years of service with men of spirit, talent and devotion. Not to mention the thrill and satisfaction flying had given me, especially in my younger days. But this was no time to be looking back; indeed, without a job to go to, my mind was now firmly focused on the future.

Part Three
Tailpiece

16 *The Aircraft Industry*

Before my resettlement course was over, I had established links
with the aircraft industry. Air Commodore Dudley Radford, the
Administrative Manager at Hawker Siddeley Aviation's (HSA's)
Manchester factory, came south looking for an aircraft sales manager,
and the Ministry of Defence (MOD) put us in touch. Dudley and
I had last met nineteen years earlier in Southern Rhodesia when he
had chaired my permanent commissioning board. However life may
have treated him meantime, his gentlemanly manner had in no way
altered and he was as calm and undermonstrative as ever. He selected
me for interview up the line, and a few weeks later I reported to
Manchester to take up post. Meantime I'd drawn less than a week's
unemployment assistance. Optimism rewarded, you could say,

My arrival coincided with a major upheaval in the sales department. Alec Watson, recruited from Dunlop as sales director, had
recently taken up post and was still in his surgeon's gown. Some
recruitment had already taken place from within the factory (Sid
Leather from the buying department and Donald Andrew from
contracts), and Brian Hough had joined the team from Hawker
Siddeley's factory at Lostock. As regards adjustments and ampu-
tations, the existing sales manager left by mutual consent before the
summer was out, and not long afterwards, we lost two more old
hands as part of a general redundancy scheme. If there *is* sentiment
in business, it rarely pokes its head above the parapet.

Recalling management changes within the factory generally, there
was no question, I had entered a new world. And one which never
quite ceased to shock me by the way in which the fortunes of

individual managers could change overnight: sometimes the sideways move; sometimes inducement to accept early retirement. The RAF, of course, had had its way of dealing with dead wood and also of clearing the way for rising stars, but the results, being spread over a huge force, were barely noticeable unless you were personally affected: for better or worse. Here, the results were focused and inescapable. So much for that aspect of my changed world.

The Manchester factories at Greengate and Woodford (originally under the proud banner of AVRO) had a distinguished record in the military field – Lancaster, Shackleton, Vulcan and Nimrod – and attitudes had been conditioned to the long production runs provided by military contracts. But by 1970 the market was changing, if local attitudes were not. Military contracts were fewer and harder to come by, and consisted of updates and role conversions. And although there were high hopes of reopening the Maritime Reconnaissance (MR)Nimrod production line through sales to countries like Canada and Australia, the production director came to rely increasingly on the HS 748 to keep his workforce fully employed.

The forty-to-fifty seater 748 twin turbo-prop airline-cum-military transport was laid down in batches against market forecasts. Alec Watson's job was to convince his fellow directors of the validity of the forecast figure and afterwards to sell those aircraft. Agonised boardroom debate increasingly became the norm as we approached decision time on the latest batch: and a decision to delay a decision was not uncommon. The market was a tough one: our chief rival, Fokker, having established something like a two-year lead with their launch of the F27 Friendship.

In many ways my years in Intelligence and Air Plans were a sound preparation for the aircraft salesman's job. Both had demanded attention to detail and meticulous follow-up action, and seldom had either produced a quick return. Selling the 748 had a lot in common. Typically, the gestation period for making an aircraft sale is something approaching two years, and to begin with I was picking up work started by others.

But if achieving an individual sale was a long haul, the tactical scene was in no way slow moving. Leaping off to Kinshasa, Rabat or Seoul at the drop of a hat and arriving in the vice-chairman's or general's office a day or so later, bright-eyed and up-to-speed on all the technical and commercial reasons why he should choose the HS748 rather than the F27, had a real challenge to it. It was also great fun to be off on one's own; out of the office, away from the telephone, winging my way over new territory and temporarily in complete charge of things. And if it could become tiring after the fourth such trip in as many weeks, there were ways of easing the strain.

Top of my list was to leave the office with a Business Class ticket on the most reputable airline flying the route: personally I love sitting behind pilots who fuss over detail and who are *not* concerned to make their destination at all costs – airfield shrouded in fog or not – just in order to save fuel. If flying East for any distance the next priority was to get a seat on the port side, out of the sun, and – whatever else happened – in the non-smoking section (sorry smokers, I sympathise with you in your problem, but I draw the line at having to share it.) With those things in place, and knowing that our agent had booked me into a reasonable hotel, I was at least giving myself the possibility of some sleep and arriving in fighting shape.

At the hotel, I refused the room alongside the lift shaft and the one over the disco too. And in certain countries quite elaborate precautions were necessary to deny Montazuma his revenge; ice-cream and salads were avoided – except perhaps on the last night – and the water I drank or cleaned my teeth with came from a freshly opened bottle, and never from a jug or the tap. Sometimes, regardless of these precautions, Montazuma triumphed, and I rummaged for the pink pills.

Almost immediately, I was despatched to Teheran, where I camped for a month trying to persuade the Imperial Iranian Air Force (IIAF) of the advantages of the 748 over the F27. We lost that particular battle, ostensibly because the latter had a better single-engine ceiling (not that the 748 was other than perfectly safe for the operating circumstances.) But, as I came increasingly to realise, favourable decisions depend on a variety of factors, and official explanations often disguise the underlying reason. The fact that the Shah had used an F27 as his personal aircraft and had taken a shine to it did not help our case. Neither, I suspect, did his close personal friendship with Prince Bernhard of the Netherlands.

We did not give up that battle in Teheran without a fight, and at one stage I was joined by Frank Wilson from the design office. His job was to help convince the IIAF C-in-C that our *projected* large side-door (openable in flight) really would work as a means of dropping paratroops and containers.

Overflying the sprawling city of Teheran at a bare thousand feet one morning, en route to a meeting with the C-in-C, our agent Hussain Zangani at the controls of his single-engined plane, Johnnie Brown the HSA regional executive alongside him, and with Frank Wilson and myself in the back, I could not help thinking that should the engine fail, our Cessna's gliding characteristics would have born a close resemblance to those of the common house brick, and that there was no way in which we were going to clear the built-up area. So much for my pernickety selection of the airline outbound from the UK.

My first success in terms of a sale came after nineteen months and, even at that, it was a follow-on order from BFS – the air calibration unit of the West German ministry of transport. It was an important success nevertheless, since it closed the door that much tighter on Fokker. In due course, and in spite of vigorous attempts by Fokker to unseat us, we added five more aircraft to that sale, making seven in all. (In the early eighties Germany was to provide us with further success when we sold six 748 aircraft to the feeder airline DLT – again in competition with Fokker – a sale in which I played an active part in getting the customer to the negotiating table, and thence to contract signature, but a prospect cultivated initially by one of my salesmen.)

When I speak of *my* success, although as the salesman I personally spearheaded the drive in pursuit of a particular sale, that drive involved a whole array of people, some in the factory, some outside: it began perhaps with a telex from our Regional Executive (resident in the area) or a tip from Harry Holmes or Rodney Hadwen of our Market Development Department, suggesting the prospect of sale;

Flying Wing, RAF Manby School of Refresher Flying *(See also lower picture facing page 251)* ABOVE. Flypast – AOC's inspection, 1967. Dominie T1s leading Jet Provost T4s. BELOW. Flypast – AOC's inspection, 1967. Varsity T1's (a poor photo of a worthy unit).

ABOVE. Air Marshal Sir Patrick Dunn taking his leave. Air Commodore 'Cyclops' Brown looks on. BELOW. College of Air Warfare dinner. Manby, 1967. ACM's Walker, Grandy (CAS), Broadhurst, Davis and Evans with the Commandant, Air Commodore Brown.

Pam Aiken's Tramp Supper, Rheindahlen, December 1969
ABOVE. Sir Christopher Foxley-Norris *centre,* with Pam Aiken *right.*
BELOW. *(left to right):* AVM John Aiken, Lady Joan Foxley-Norris and
Isabel Hazelwood.

Harrier hovering before touchdown alongside Rheindahlen Combined HQ.

RIGHT. Sir Denis Spotswood, C in C RAF Germany watching the Harrier's approach.

Some of my sales team's HS/BAe 748 successes
opposite page: BELOW. West Germany *(radio calibration unit). this page:*
ABOVE. South Korea *(aircraft in foreground.)* BELOW. West Germany
(feeder airline). (Photos courtesy of Avro International – Ian Lowe).

ABOVE. Nimrod MR2 in Falklands trim *(Photo courtesy of Avro International – Ian Lowe).* BELOW. Aerospace Show, Iruma, Japan, 1973. Author explaining Nimrod MR2 to Mr Nakasone *facing camera,* Minister for International Trade and Industry.

it involved not only the support of Eric Johnson and his sales engineers, but occasionally (where substantially new design was required) the direct participation of design office staff; I consulted regularly with Defence Sales and the Department of Trade and Industry (DTI), London based HMG organisations in touch with out attachés abroad. In the field, our regional executives and agents apart, I drew on the British Embassy from the ambassador downwards; and at the stage of contract negotiation, quite naturally, the lead role in detailed discussion fell to one of our contracts officers. But the fact remains, the salesman cannot relax until the contract is signed. When a sale is lost, no one clamours to be identified with his failure. At the end of the day, it tends to be *our* success or *his* failure.

Contract negotiations can be gruelling affairs. One I recall went on for five weeks before the customer finally signed a contract for two aircraft. Meantime – in addition to a fundamental difference between us over price – he, or rather his pilots and technical staff, continued to haggle over detail such as the exact positioning of individual instruments on the pilot's flight panel, so that each evening we presented yet another message to the hotel telex operator for overnight despatch to our design office. Quite soon we were also acquainted with the telex operator's view of *her* priorities. Put simply, unless her palm was greased our telex stayed at the bottom of the pile. After she had woken me for the second consecutive night at one o'clock to say she was having difficulty contacting Manchester, the penny dropped. The remedy applied, the lines to Manchester gave no further trouble.

In every contract negotiation, there comes an unpleasant period when the customer applies the final turn of the screw, just to satisfy himself that he really has squeezed the last drop of advantage from the deal; threats of phoning the factory direct on the point at issue or of breaking off negotiations being not uncommon.

In one instance I remember the customer's spokesman refusing to talk any further with our contracts officer and insisting on negotiating from there on through our lawyer, and doing so largely in his (the customer's) native tongue. A little later in the proceedings, he suddenly leapt to his feet and, without as much as by your leave, opened wide all the windows before stalking out of the room. Since the temperature outside was close to freezing, his action could only be designed to add to our discomfort. He was away for ten minutes, whilst we got steadily colder. On his return he closed the windows and we resumed our talks: his mood now, one approaching affability! I wondered how long it had been since he had left the interrogation branch of his country's armed forces: his technique Pavlovian to the letter.

Since this sticky phase of negotiations is often followed within twenty-four hours by contract signature and a champagne celebration laid on by the customer, in retrospect such tactics appear unbelievably crude. But at the time they are real enough, as you face the possibility of everything going down the drain.

Now and then, doing business with authoritarian regimes, one got a hint of the insecurity felt by those at the top in such countries. Almost my first port of call on arriving in — was to present my credentials to the head of the intelligence and security services. Entering his office, a gaunt, tense-looking figure dressed in a dark grey suit rose from behind the desk, the three buttons on his single breasted jacket each carefully fastened, giving him an awkward, trussed-up appearance. In that moment, as we shook hands, I supposed that either he personally lacked dress sense or that the — tailors had yet to catch up with the Savile Row look. Only on reflection did I realise he was probably concealing a bullet-proof waistcoat. Some years later that same official was assassinated. Had he grown careless, I wonder?

In the matter of demonstration flights, we were supported by a number of excellent pilots. More than that – certain of them also radiated a natural charm and enthusiasm. None more so than Charles Masefield. What is more, when it came to demonstrations Charles could come up with some quite innovative ideas, as I discovered in Khartoum.

I explained to Charles the job to be done – to demonstrate the aerial delivery of supplies and the dropping of paratroopers, and somewhere along the line to convince the customer of the 748's rough-field capability. Charles responded with a brilliantly simple idea: I should position myself, the members of the evaluation committee and a team of paratroopers on a nice piece of flat desert a few miles out of Khartoum; he would begin by dropping a string of containers, then land and take off on the virgin desert (emplaning the paratroopers in between) and finish his act with the para-drop. In accepting the idea I practically snatched his hand off.

To avoid the strong winds prevalent later in the day, the Sudanese Brigadier and I were in position at dawn: the paratroopers at hand. As the first rays of light flooded the desert it was like a scene from a Hollywood film – silhouetted on the skyline to one side a lengthy camel train loping gently along, and on the other a string of horses being exercised. It was a magical moment. And from a practical point of view the beautiful stillness of the air made it ideal for the para-drop.

Charles appeared, smack on time, and dropped the containers. By now the paratroopers were out of their truck, lined up and ready to emplane: their pensive ebony faces glistening with a light sweat, as much perhaps with apprehension at the thought of jumping from a strange aircraft as from the weight of their equipment. The brigadier looked pleased with the supply drop, which had run nicely along the line of the Dropping Zone, and fixed his attention on the aircraft as it made an impressive short landing within a few hundred yards from where we stood. At this point I restrained him from moving forward, pointing out that the aircraft would come to us. But events were no longer obeying the script.

Even as I spoke, the aircraft came to a standstill and the engines were shut down. Apart from the fact that Charles was to have taxied to where we stood, the plan was to keep the engines running as the troops embarked. Something was badly amiss. Charles – who for once was not grinning broadly – arrived on foot to explain that a loss of hydraulic fluid had robbed him of this brakes, and that until more fluid could be obtained the aircraft would have to remain where it was. So much for the stress I had laid on the 748's reliability. Inwardly mortified, I set about picking up the pieces with the evaluation team, leaving the aircraft to be rectified and repositioned at the airfield.

Twenty-four hours later, the Brigadier and I were standing on the same spot watching his paratroopers float gently down to land within yards of us; their faces on this occasion positively wreathed in smiles as they gathered up their parachutes and made their way to their truck. Meanwhile, the Brigadier had been very understanding – not every officer of that rank would have been willing to get out of bed two mornings in a row at five o'clock just to please a manufacturer. Having fallen on our faces the morning before, I could only hope that we had impressed with our speed of recovery, and that the Brigadier had accepted my word that such a failure was quite exceptional. As to the speed of recovery – all credit to Charles and his support engineer Frank Lord.

Resourcefulness too is sometimes a vital quality in demonstration pilots. In Zaïre, Eric Franklin showed it in plenty when he rescued us from a potentially embarrassing situation. At the request of the senior member of the evaluation team, we had landed at an earth strip well out in the bush. The landing (a deliberately short one) had presented no problem, but Eric wanted to inspect the far end of the strip more closely on foot before taking off, to see how much his run might be restricted by a soggy looking patch. I decided to accompany him; leaving our agent to hold the fort with the evalluation team.

In spite of an explicit request that everyone would remain with the aircraft, we returned from our inspection to find that most of the team had disappeared to the local village – the evaluation team leader's birthplace, would you believe! Meantime the aircraft had begun to sink into the unmetalled surface. Problem – with no transport available, how to retrieve our passengers from a village a mile or so away before the aircraft became completely bogged down? The answer, provided by Eric, was to take off and remain airborne until our wanderers returned. The sound of his departure had them back in no time.

In my early days at Manchester, getting action on prices and deliveries after an overseas visit seemed to take an age. The factory system and its devotees thought in weeks rather than days – a legacy no doubt of long production runs, when time was rarely of the essence. But, thankfully, organisations are composed of individuals, and I soon learnt who to turn to in order to speed things up. Frank Wilson, assistant chief designer, could produce a document (for pricing purposes) describing a special requirement practically overnight once he was convinced of the need, and do so with accuracy and flair. On the same basis, I found allies in the contracts department too. Latterly, by imposing priorities, the basic system shed some of its lethargy. But, to the end, there remained the need to chivvy that system and frequently to fight one's corner as priorities shifted.

It would be unfair not to redress that ponderous image of BAe Manchester by citing what it *could* do when it put its mind to it. As during the Falklands crisis, when the MOD called for RAF Nimrods to be given an in-flight refuelling capability. The timescale – yesterday. Eighteen days from the first MOD telephone call, the first Nimrod had been converted and was returning to its unit, hooking up to a tanker aircraft en route to complete the final stage of testing. That emergency situation revealed Manchester's true metal, when people like Alan Clegg, chief designer Nimrod, came up with a brilliantly simple time-saving solution for the internal fuel piping system, and under Vernon Horsefield, the works manager, the shop floor worked night and day to complete the job.

Some of my most interesting visits were spent in pursuit of co-production projects in the company of people from other disciplines – commercial, design and aerodynamics – Stephen Ward, Alan Troughton and John Richardson are among the names which spring to mind. Chiefly we were promoting a maritime version of the 748. Apart from the satisfaction of matching our wits against those sitting across the table – usually stimulating enough in itself – such visits sometimes threw up the bonus of seeing some well known feature of historical interest, and just occasionally we were confronted with something entirely unexpected but equally fascinating.

In the latter category were the enormous Zeppelin hangars on the French Aeronvale airfield at Cuers, near Toulon; part of the First

World War reparations extracted from Germany. How very empty and incongruous those massive hangars appeared; like two huge dinosaurs which had refused to die off. But most of those co-production-seeking visits were to India, which gave me the opportunity to re-acquaint myself with Delhi and Bangalore, as well as to make up for the frustration of thirty years before when fog had denied me even an aerial glimpse of the Taj Mahal. Visiting Agra over a weekend I was at last able to inspect it, and to do so at ground level and at my leisure. What an air of tranquility that noble building exudes – encased in the purest of white marble, the surface relieved with beautiful mosaics in cornelian, jasper, agate and lapis lazuli; the interior decorated with marble screens, pierced and sculptured like intricate lacework.

To return to business, and our quest to interest India in the co-production of a maritime version of the HS748: in Delhi the wheels of government grind so slowly and the cup is so often dashed at the last moment, that one should never be surprised when commercial success remains, once more, around the corner. Strong men have been known to return from India and to go bed for a week exhausted by the frustration of it all. To put it briefly – in spite of our established foothold with the 748 in India – we failed. The politics behind our failure would make interesting reading. But I think I should leave it at that.

On the surface, a replacement trainer for the French navy – based on the 748 Coastguarder – looked to be a project of quite a different kidney as we progressed smoothly towards a final contract. We had negotiated the split of production at the government factory at Cuers and completed discussion of the second draft of a contract with officials in Paris when the rug was unceremoniously pulled from under us. By British standards, our treatment was unbelievably cavalier, and the French officials themselves were clearly embarrassed by this eleventh-hour pull out. Out of the blue, the Nord 262 was retrieved from among the rejected solutions and substituted for the 748. Even the representations of Mr Francis Pym, our Foreign Secretary, could not restore our position; such was the power and influence of Monsieur Dassault – aircraft manufacturer and parliamentary deputy. Whether an ill-timed article in the *Flight* magazine foolishly gloating over our imminent success, had anything to do with the turnabout it's hard to say, but had I been M Dassault, I would have seized upon it as heaven-sent ammunition with which to stir up nationalist sentiment in support of my case.

By the early seventies, in parallel with the sales department's 748 activities, an ad hoc team was formed to sell the MR Nimrod to a number of government-approved countries. The immediate target was the Royal Canadian Air Force, closely followed by the Royal Australian Air Force (RAAF). As an entirely military product, however, and following past custom, this selling venture was led by the design office, with Humphrey Wood, our General Manager, directing the major moves and personally leading the team in the field. Considerable effort went into these activities and it was refreshing to see Humphrey, our top man, being prepared to stake his personal reputation on the outcome. Nevertheless, we lost the competition to Lockheed in both countries. Canada ordering the CP-140 Aurora (basically a P3C) and Australia the Orion P3C.

As to the reasons for those failures: on the face of it, perhaps the overriding factor in the case of Canada was the sheer volume of offset work promised by the Americans to Canadian industry; and with the RAAF already operating the P3B (an earlier version of the Orion) and Australian industry having tasted the benefits of the P3B deal, it was something of a tall order to expect that we could dislodge Lockheed. No doubt our people directly involved with those projects might cite additional reasons.

With those successes already to Lockheed's credit, we went on to oppose them and their P3C in Japan and the Netherlands: both already operating Lockheed aircraft (Neptunes) in the maritime role. By late 1973, I was acting as Nimrod sales manager and spent some time in both countries.

The RAF, with Defence Sales backing, gave our efforts in Japan excellent support. They flew a Nimrod to the 1973 Iruma airshow and – in addition to participating in the airshow itself – they allowed its crew to give our contacts in the Japanese navy a series of demonstration flights. (I must admit, given my wartime background, it felt a trifle odd to be flying over Tokyo Bay with a Japanese pilot at the controls of an aircraft boasting RAF roundels.) Reverting to the quality of RAF support; with the show due to open the following day, the first Nimrod to arrive at Iruma suffered partial engine failure during rehearsal. But such was the RAF's degree of commitment that within twenty-four hours a replacement aircraft had arrived to see the job through.

Similarly, our efforts to sell the Nimrod to the Royal Netherlands Navy (RNN) were generously supported by the British government, both in the guise of the Procurement Executive – in allowing us to use one of their Nimrods for demonstration flights – and by our embassy in the Hague, whose help we enlisted both to persuade the RNN to accept a sales presentation and in arranging for us to address one of the Dutch parliamentary committees.

But in both instances our efforts ended in failure, and again the P3C won the day. Whereas we knew it would not be easy to dislodge Lockheed, we were left with the feeling that the decisions had not been taken without some partiality. Indeed, the Japanese prime minister of the day subsequently faced charges of accepting payments from Lockheed, and a highly placed person in the Netherlands had his relationship with Lockheed called into question in the Dutch Parliament. Sour grapes? Not entirely, I think.

It was by no means my final fling with Nimrod. Several years later I was a member of a combined British Aerospace and GEC team which made several visits overseas attempting to sell the Airborne Early Warning (AEW) version. One country in particular seemed desperately keen to acquire an AEW capability, and we really thought we were on to something. But the officer in charge of procurement – who had an engineering degree and who probed technical areas in some detail – refused to accept GEC's claims for the system's avionics, and the prospect withered. The subsequent RAF decision to abandon Nimrod AEW in favour of the Boeing E3A AWAC aircraft – although taken on the grounds that the engineering of the AEW avionics had not reached an acceptable standard, as opposed to faulting the underlying design of that system – must, nevertheless, have given that officer some cause for satisfaction in not proceeding. Not that that was of any comfort to British Aerospace.

As the salesman in these AEW excursions, it fell to me both to set up the appointments and to make the administrative arrangements, so I had a dual interest in how things went. Naturally, the thing which mattered was a successful outcome to our meeting with the customer. But occasionally, given the need to meet a particularly tight schedule, there was some satisfaction to be had purely from seeing one's plans hold together. An AEW trip to Australia (in October 1983) with John Scott-Wilson, our technical director, was a case in point.

My brief was to set up meetings in Singapore, Canberra and Kuala Lumpur so that John was out of his office for no more than one working week; the meetings to cover prospective users, their associated procurement agencies and the respective British High Commissions. We left Manchester airport on Saturday and, having landed back at Manchester on Sunday eight days later, we were in our offices first thing the next morning; all meetings held as planned. Not a schedule one could set up very often, let alone survive on a regular basis, but very satisfying as a one-off achievement. On our last night in Australia – the job done – I had even managed to arrange for Terry Malcolm (my wartime wireless operator) and his wife to join us for dinner. Something I could not have achieved without John Scott-Wilson's encouragement and some rapid detective work on the part of our agent Nat Gould in tracing Terry's phone number. As I hadn't seen Terry for thirty-eight years, we had plenty to talk about.

Looking back at the government's nationalisation and merger of Hawker Siddeley Aviation (HSA) and the British Aircraft Corporation (BAC) to form British Aerospace (BAe) in April 1977, this hardly touched the lives of individuals in the industry in the immediate sense. But no one was quite certain what to expect in terms of management redundancies in the longer term, and events at the local level were not without a touch of irony. With the formal announcement that the merger was to take place, our general manager promptly issued a note to his executives asking for their continued

loyalty under the new arrangement, only to announce his own departure (to accept an offer he 'could not refuse') a week or so later. An
event received with some dismay and disillusionment by his loyal
band. By a strange coincidence, the commercial director also chose
that moment to pack his bag for a private company. So much for
loyalty.

Having dwelt somewhat on our failures in the field of military
sales, I can see that I am in danger of leaving my reader with a
rather negative view of events at HSA/BAe Manchester during the
nineteen-seventies and early eighties. That is far from being a balanced picture. For a good deal of that period the men who had
previously built the maritime Nimrods were engaged in converting
Victor bombers to the tanker role and updating the avionics in the
RAF's maritime Nimrod fleet. And with the 748, we had our sales
successes too – civil and military – which kept that production line
going. We were also adapting to meet future requirements: under
Alan Troughton's design leadership a new aircraft was being
developed to succeed the ageing 748 – the ATP (Advanced Turbo-
Prop), an aircraft with greater seating capacity, more fuel-efficient
engines and updated systems; the Chadderton factory began the
manufacture of rear fuselages for the jet-engined BAe 146 and took
on an increasing volume of Airbus work; and plans were also afoot
to lay down a BAe 146 assembly line at Woodford.

In early 1986, hard on the heels of a visit to Pakistan, I was
whisked into hospital by ambulance with rigors and mental disorien-
tation. There, I shot a temperature of 107 plus. Suddenly the room
in which I was isolated became a scene of hyperactivity as a nurse
sponged me down with iced water while two doctors worked at
hectic speed; one taking a blood sample; the other loading a syringe
with antibiotics with which to follow up. The hospital seemed unsure
of what had triggered those alarming symptoms but suggested a
probable infection of the bladder, causing septicaemia. Considering
the brittle condition of my prostrate after previous bouts of surgery,
that seemed plausible enough, and I could only conclude that Pakistan had been the source of infection.

A week later I was discharged from hospital. But full recovery was a slow and, at times, distressing business, what with persistent insomnia, periods of unaccountable anxiety and a feeling of general weakness – so much so that I decided to ask for early retirement. To carry on with my job – essentially one of overseas travel – and risk another such infection, seemed a gamble hardly worth taking at the age of sixty-two, given an acceptable alternative. Charles Masefield, by then our general manager, gave my application the necessary push.

So ended my days in aviation. A career of over forty years. The wartime ones characterised by heightened risk and an atmosphere of close comradeship; the post-war period offering the satisfaction of instructional duties and the stimulus of jet flying; the latter years of my RAF service providing the opportunity of command and requiring an increasing use of the pen. And, finally, in the aircraft industry: learning to sit in the *back* of an aircraft; being part of an organisation making strenuous efforts to adapt; working with individuals every bit as enthusiastic and professional as those in light blue and, most important, selling some aircraft. That wet-behind-the-ears RAF medical officer who gave me the benefit of the doubt in January 1941 had indeed opened a door.

APPENDIX A
Personal Record of RAF Service

Wartime

Jan 42	No. 1 Aircrew Receiving Centre, Lord's Cricket Ground, London
Mar 42	No. 6 Initial Training Wing, Aberystwyth
May 42	No. 6 Elementary Flying Training School (EFTS), Sywell
Sep 42	No. 17 EFTS, Stanley Nova Scotia
Nov 42	No. 8 Service Flying Training School, Moncton New Brunswick
May 43	No. 6 EFTS, Sywell
Jun 43	No. 6 Advanced Flying Unit, Little Rissington
Jul 43	No. 1516 Beam Approach Training Flight, Pershore
Sep 43	No. 15 Operational Training Unit, Harwell
Jan 44	No. 1584 Heavy Conversion Unit Salbani/Kola India
Jul 44	Liberator Captain, No. 356 Heavy Bomber Squadron, Salbani
Aug 45	Operations Instructor, Liberator Refresher Flying Unit, Kolar
Dec 45	PA to Air Officer Commanding 225 Group, India
Nov 46	Left RAF at end of wartime emergency period

Postwar

Aug 48	Supernumary Pilot, No. 83 RAFVR Centre, Stanmore
Apr 49	Rejoined RAF, Waterbeach
May 49	Pilot Refresher Flying Unit, Finningley
Jun 49	Adjutant, No. 666 Air Observation Post (AOP) Squadron,

	Perth
Jul 49	Trainee Flying Instructor, Central Flying School (CFS)
Mar 50	QFI and Flight Commander, No. 4 FTS, Heany Southern Rhodesia.
Dec 52	Qualified Flying Instructor (QFI) and Flt. Cdr CFS
Nov 54	Flt Cdr No. 25 All-Weather Fighter Squadron, West Malling
Oct 55	Flt Cdr No. 219 All-Weather Fighter Squadron, Driffield
Dec 56	Station Commander, No. 1 Signals Unit, St Margaret's Bay
May 58	OC Advanced Standards CFS Little Rissington
June 60	AI3(i) and AI3(a) Air Ministry – Air Intelligence
Mar 64	Student, 29 Course, Joint Services Staff College, Latimer
Sep 65	Station Commander and OC School of Refresher Flying RAF Manby
Nov 67	Group Captain Air Plans Headquarters RAF Germany
Feb 70	Left RAF on voluntary retirement

APPENDIX B
Aircraft Flown

As Qualified Pilot

Piston Engine
DH82a Tiger Moth
Fleet Finch II
Anson II, 19 and 21
Oxford 1
Wellington 1c
Liberator III, VI and X
Auster V and VI
Harvard IIA, IIB, III & T6G
Spitfire XVI
Prentice T1
Mosquito TIII
Chipmunk T10
Provost T1
Pioneer (S. Eng) CC1
Balliol T2
Varsity T1

Jet
Meteor T7, F4, F8, NF12 & N F14
Vampire FB5, T11 and T22
Venom NF2 and FAW21
Hunter F4, T7 and T8
Canberra T4 and PR3
Jet Provost T1, T3 and T4

Sabre 6
Seahawk FGA
Dominie T1

Other Aircraft Flown

Lancaster VII
Lightning T4
Piaggio P149D
Fouga Magister CM170 R

APPENDIX C
Abbreviations and Special Terms

A & AEE Aircraft and Armament Experimental Establishment
ACM   Air Chief Marshal
AFS   Advanced Flying School
AOC   Air Officer Commanding
AOC in C Air Officer Commanding in Chief
ATC   Air Training Corps
ATP   Advanced Turbo-Prop
AVM   Air Vice Marshal
AWF   All-Weather Fighter
BAe   British Aerospace
balbo   Large formation of aircraft
BAOR   British Army of the Rhine
CAS   Chief of Air Staff
C in C   Commander in Chief
CFS   Central Flying School
cu/nimb cumulo-nimbus cloud
DTI   Department of Trade and Industry
EFTS   Elementary Flying Training School
ETPS   Empire Test Pilots' School
ETA   Estimated Time of Arrival
FB   Fighter Bomber
FTS   Flying Training School
g   Force exerted by gravity
gaggle   A loose collection of aircraft
GCA   Ground Controlled Approach
GCI   Ground Controlled Interception

HSA	Hawker Siddley Aviation
i/c	In Charge
IP	Initial Point
ITW	Initial Training Wing
LAC	Leading Aircraftman
Met.	Meteorology
MOD	Ministry of Defence
MR	Maritime Reconnaissance
NF	Night Fighter
OC	Officer Commanding
OTU	Operational Training Unit
PA	Personal Assistant
PI	Practice Interception
PSO	Personal Staff Officer
QFI	Qualified Flying Instructor
RAFG	Royal Air Force Germany
RATG	Rhodesian Air Training Group
RCAF	Royal Canadian Air Force
RCeyAF	Royal Ceylon Air Force
SAF	Strategic Air Force
SFTS	Service Flying Training School
stick	Aircraft control column
stick	A group of bombs released to straddle a target
Wings	Pilot's flying badge
Wop/AG	Wireless Operator/Air Gunner

INDEX